A Guide to the Four Most Common Species of Root-Knot Nematodes (*Meloidogyne* Spp.), With a Pictorial Key

J. D. Eisenback
H. Hirschmann
J. N. Sasser
A. C. Triantaphyllou

A Cooperative Publication
of
The Departments of Plant Pathology and Genetics
North Carolina State University
and
The United States Agency for
International Development
Raleigh, North Carolina
May, 1981

For information on the availability of this publication, write to:

J. D. Eisenback, Ph.D.
3510 Indian Meadow Drive
Blacksburg, VA 24060

First Printing of Revised Edition, 2023
(Softcover)

ISBN 978-1-893961-00-5

Library of Congress Cataloging-in Publication Data
Names: Eisenback, Jonathan David, 1952- ; Hedwig Hirschmann Triantaphyllou, 1927-2021; Anastasio Christus Triantaphyllou, 1926-2009; Joseph Neal Sasser, 1921-2005.
Title: A Guide to the Four Most Common Species of Root-Knot Nematodes (*Meloidogyne* Species), with a Pictorial Key: Mactode Publications [2023]
Identifiers: ISBN 978-1-893961-60-9 (hardback)
Subjects: LCSH: Nematodes, Nematology

Author: Jonathan D. Eisenback
Hedwig Hirschmann
Joseph N. Sasser
Anastasio C. Triantaphyllou
Copy Editor: Marilyn L. Eisenback
Art Editor and Designer: Jonathan D. Eisenback

Second Edition, 2023 Printed in the United States by Kindle Direct Publishing and Published by Mactode Publications Blacksburg, VA 24060

PREFACE

Control of root-knot nematodes, *Meloidogyne* spp., by crop rotation or through the use of resistant cultivars can be accomplished only if the species and host races to be controlled are known. Accurate identifications are necessary because some species or races attack certain crop plants, whereas others do not, and resistance developed in one crop cultivar is not necessarily effective against all species or races of root-knot nematodes. Because of the importance of identification in the design of effective control programs, there is need for a rapid and reliable method to identify populations of root-knot nematodes.

To date, approximately 45 species of *Meloidogyne* have been described and new ones are being reported each year. Although most species are adequately described, it is becoming increasingly difficult to distinguish between species on the basis of presently available information. Difficulties arise from the considerable variation among individuals of a species in many of the characters now used to distinguish between the species. For example, perineal patterns are quite variable, and measurements of various other characters often overlap between species.

A major objective of the International *Meloidogyne* Project (IMP) has been to reevaluate the various types of taxonomic characters (morphology, host response, cytology, and biochemistry) and to discover new and more reliable characters useful for species differentiation. We have looked at many populations in detail by scanning electron microscopy (SEM) and light microscopy. In addition, plant response to the various species and races using certain host differentials has been added to supplement the morphological data. Hundreds of populations have been characterized cytologically to determine the mode of reproduction and number of chromosomes for each species. Some diagnostic information is also available from biochemical studies adding additional characters for species identification. Although it may be possible to identify a population by observing a limited number of characters from a few specimens, more often it is not, particularly for investigators with limited experience in *Meloidogyne* systematics. Furthermore, placing too much emphasis on one or two characters, however reliable they might appear, increases the chance of an incorrect identification.

The major purpose of this guide is to assist the cooperators of the IMP, as well as other nematologists, plant pathologists, and other investigators, with the identification of the four common root-knot nematode species, namely *Meloidogyne incognita*, *M. javanica*, *M. arenaria*, and *M. hapla*. These are economically the most important species and are responsible for 90% or more of the damage to agricultural crops caused by this genus. As sufficient information becomes available for other species or groups of species, supplements to this publication may be prepared.

This guide emphasizes identifications based on morphological and host-response data which can be easily obtained by investigators with limited experience and laboratory facilities. Useful supplementary data on fine structure as revealed by SEM, cytogenetics, and biochemistry are also provided. Unfortunately, these supplemental characteristics require trained personnel and sophisticated equipment and facilities which may not be available in many laboratories.

The authors wish to thank Miss Milly Oldham for typing this manuscript. Thanks are also extended to Dr. K. R. Barker, Dr. J. L. Starr, Mr. A. L. Taylor, and Miss Gean Cliff for reviewing this publication and making many helpful suggestions. We especially wish to thank the United States Agency for International Development (USAID) for supporting the International *Meloidogyne* Project and for making copies of this guide available to our cooperators in developing countries.

PREFACE to the Second Edition

Although the first printing of this book was more than 40 years ago it is still relavent today since most populations of root-knot nematodes can easily be identified from their morphology. During the 40 years between printings, the number of described species has more than doubled, but the four most common species remain the most important. A few species, like *M. chitwoodi, M. enterolobii, M. naasi*, and *M. graminicola* are becoming more important, probably because they are more often identified correctly.

Molecular techniques have all but replaced morphological ones; however, morphology remains much faster and cheaper. As it turns out, molecular techinques have a difficult time separating out the three common species of *M. incognita, M. arenaria*, and *M. javanica* and specific tests are necessary to identify them.

This second edition has been enhanced with a hardcover and colored type. Where possible, original photographs were re-scanned. In addition, other enhancements have been made including the correction of a few errors in layout and typography. Hopefully, more corrections were made than new errors were introduced.

JDE

CONTENTS

Part I. A General Description of Several Taxonomic Approaches to *Meloidogyne* Species

I. Morphology

The basic morphology of *Meloidogyne* species is quite similar. Nevertheless, certain distinguishing characters are useful in species differentiation. These characters include the morphology of perineal patterns, the head morphology of females, males, and second-stage juveniles, and the stylet morphology of females and males **(Figs. 1.1-1.8)**. Perineal patterns and head shapes of males appear to be the most helpful characters. Stylet morphology is also reliable but can be used only in specimens that are properly prepared and viewed in exact lateral position. Additional characters, such as number of lines in the lateral field, may be useful in the identification of some species and will be listed where appropriate. Also, certain morphometric data which may be helpful in eliminating some species from consideration will be listed. A population should never be identified from measurements alone, however, because they are variable and overlap between species.

Details of the most useful characters are shown in scanning electron microscope (SEM) photographs and light microscope (LM) photographs and drawings for each species in **Part 3** of this guide. Presently it is not possible to routinely identify species with the SEM as such an instrument is not readily available to many workers. The SEM, however, greatly clarifies some of the morphological details that are visible in the LM, and SEM photographs are thus included here to help interpret characters seen in the LM.

The morphology of more than 100 populations of the four common species from different geographical areas around the world has been examined. Some morphological variation was observed between and within populations of these species, but most populations can be readily identified when several characters are considered. We have attempted to select for each species what we believe to be a "typical" population. Most of the photographs and drawings presented are therefore based on a single population of the particular species selected from the live *Meloidogyne* collection at North Carolina State University.

1.1

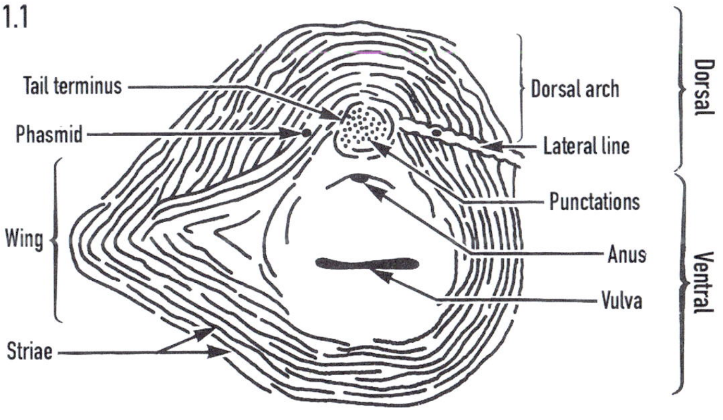

Fig. 1.1. Generalized morphology of a perineal pattern

1.2

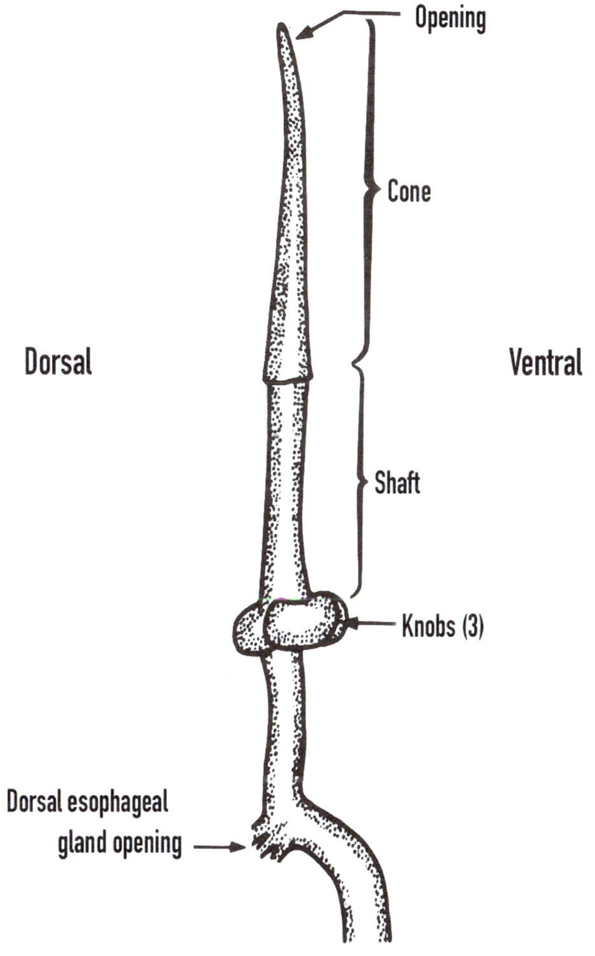

Fig. 1.2. Basic morphology of a stylet of a root knot nematode female

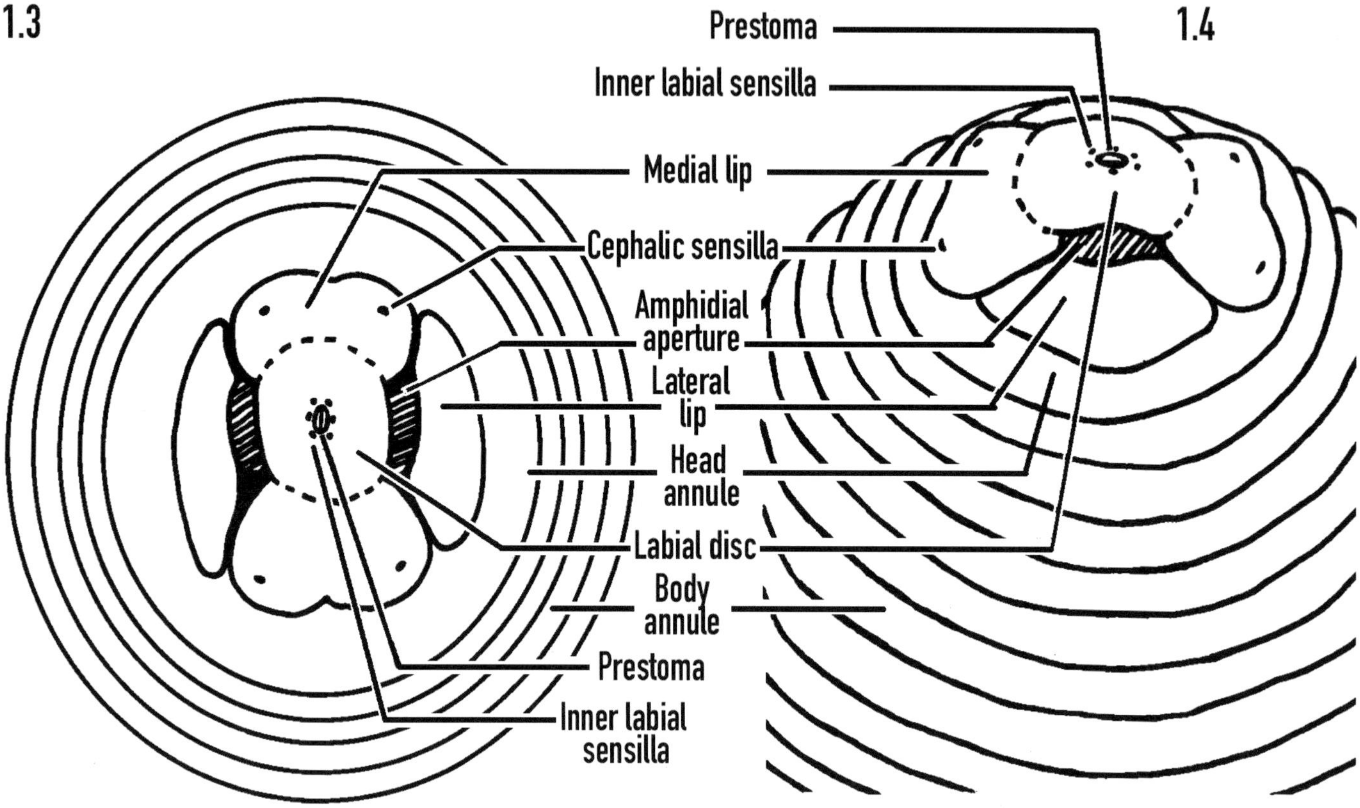

Figs. 1.3-1.4. Diagrams illustrating the generalized head morphology of a female as revealed by SEM. 1.3) Face view. 1.4) View from the lateral side

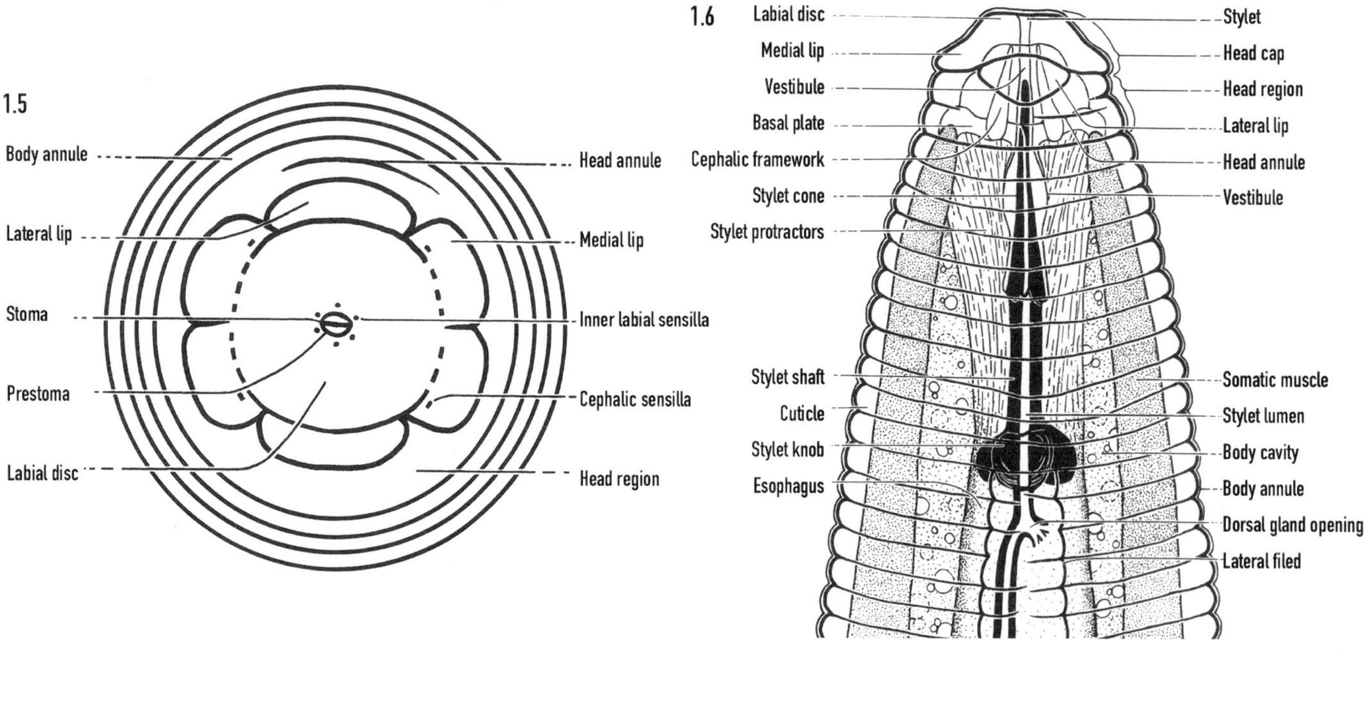

Figs. 1.5-1.6. Illustrations of the basic morphology of the head and stylet of a male as determined by LM and SEM observations. 1.5) Face view (SEM). 1.6) View from the side (LM and SEM)

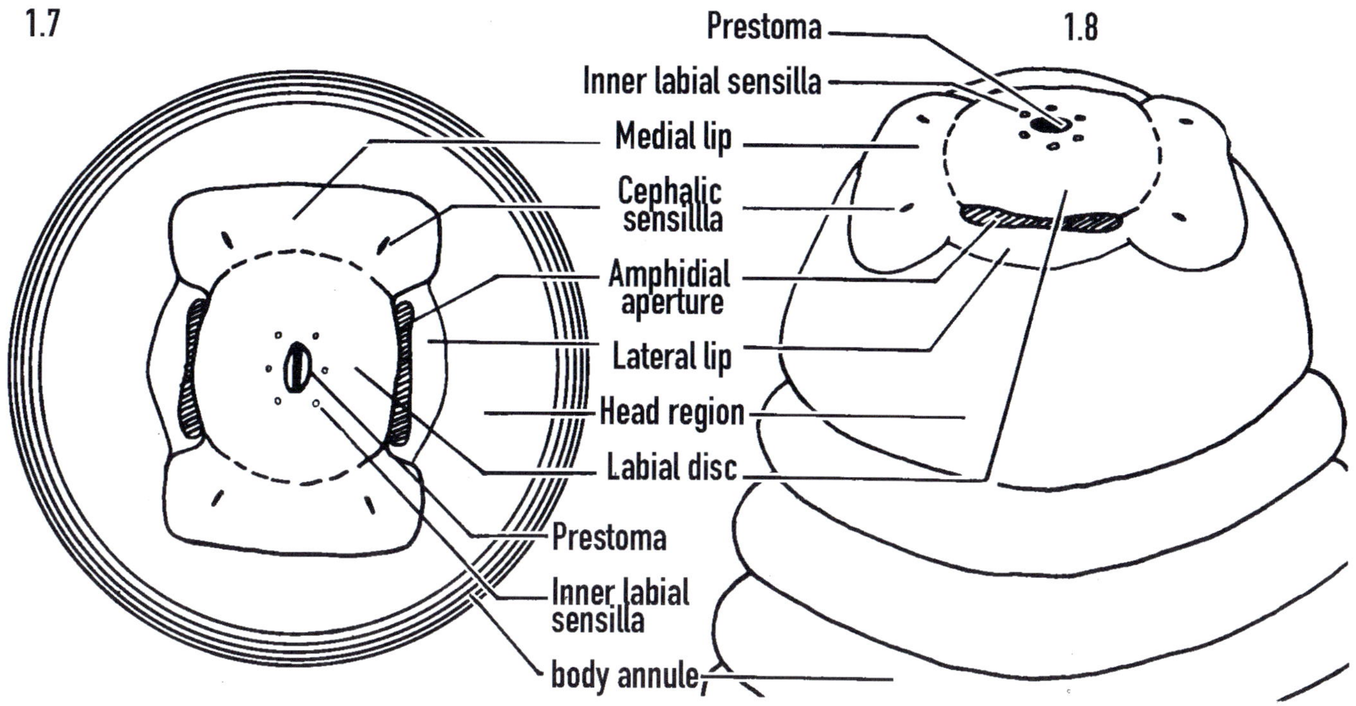

Figs. 1.7-1.8. Generalized head morphology of a second-stage juvenile as revealed by SEM. 1.7) Face view. 1.8) Lateral view

II. Differential Host Test

The differential host response test gives a preliminary indication of the root-knot nematode species in question based on the typical host response **(Table 1.1)** and can detect parasitic variation as evidenced by host responses significantly different from the usual responses.

The differential host test cannot be relied upon entirely for identification because the population may contain more than one species or the population may comprise a species for which there is limited or no host response data. The differential host response test is fairly reliable for identification of the four common species.

Appreciable deviation of a population within a species from the usual reactions indicates that the population under consideration is different from the norm. For instance a few populations of *M. javanica* have been detected which infect and reproduce on pepper and peanut. Most populations of *M. javanica* do not infect or reproduce on these two crops. Thus through the differential host response tests, involving hundreds of populations of the different species from many parts of the world, useful information can be obtained concerning possible differences in parasitic behavior between populations of a given species. Final identifications, however, should be based on morphological, cytological, and biochemical data.

Table 1.1. Usual response of the differential hosts to the four common *Meloidogyne* species

Meloidogyne species	Tobacco	Cotton	Pepper	Watermelon	Peanut	Tomato
M. incognita	-* (+)	+*(-)	+	+	-	+
M. javanica	+	-	-* (+)	+	-(+)	+
M. hapla	+	-	-	-	+*	+
M. arenaria	+	-	+	+	+*(-)	+

[1] Plant varieties include: Tobacco, NC95; Cotton, Deltapine 16; Pepper, California Wonder; Watermelon, Charleston Gray; Peanut, Florrunner; Tomato, Rutgers.
[2] ASTERISK indicates key differentials for that species.
[3] Parentheses indicate that a small proportion of the populations attack that host.

III. Symptomatology

The above ground symptoms of plants infected with root-knot nematodes are similar to those caused by other root pathogens and/or environmental conditions that restrict the uptake of water or nutrients. Symptoms such as reduced growth, chlorosis of foliage, susceptibility to wilting, and reduced fruit production are most common. The most characteristic symptoms of infection by root-knot nematodes occur on the roots. Most *Meloidogyne* species cause the infected root to swell around the feeding nematode and the typical root gall is formed. Galls may occur singly or several galls may coalesce to form very massive galls. Some species also stimulate the plant to produce many lateral roots which emerge from the gall and result in a thick, matted root system. Even though some species produce a characteristic type of galling, species identifications cannot be based solely on these root symptoms.

IV. Cytogenetics

Cytological and cytogenetic information can be used to supplement morphological data in species identification and often to verify identification of certain major species. The most important cytogenetic characters of root-knot nematodes are: mode of reproduction, process of maturation of oocytes, and chromosome numbers **(Fig. 1.9)**. Some species reproduce by cross-fertilization (amphimixis), others by parthenogenesis (obligatory mitotic parthenogenesis), and still others by cross fertilization and parthenogenesis (facultative meiotic parthenogenesis). Cross-fertilizing and facultatively parthenogenetic species undergo meiosis during maturation of oocytes and this involves pairing of homologous chromosomes and formation of bivalents (tetrads). The haploid number (n) of bivalents is observed at metaphase of the first maturation division in these species. Obligatorily parthenogenetic species do not undergo meiosis and maturation of oocytes in such species consists of a single mitotic division. The diploid number of univalent chromosomes (dyads) is observed at metaphase of the single maturation division. In addition to these differences, the chromosome number may be different in various species and also may vary within species. In general the haploid chromosome number in species undergoing meiosis varies from n=14 to 19. The diploid chromosome number observed in ameiotic species varies from 2n=30 to 56.

These cytogenetic features are useful taxonomic characters and can be very helpful and reliable in identification of some species. Although obtaining such cytogenetic data requires some familiarity with cytological procedures and availability of certain laboratory facilities, it would be beneficial for research laboratories with major interest in root knot nematodes to have these capabilities.

V. Biochemistry

Analysis of proteins via polyacrylamide gel electrophoresis can supply additional information for distinguishing species of root-knot nematodes. Electrophoretic patterns of several enzymes, particularly esterases, malate dehydrogenase, and a glycerophosphate dehydrogenase, are different for each species.

Esterase patterns are the most useful for identification of the four most common species of *Meloidogyne* and are included in this guide **(Fig. 1.10)**. The relative electrophoretic mobility (Rf) of the major esterases varies according to the electrophoretic conditions employed in each laboratory. For this reason we have standardized the Rf of the major band of *M. hapla* to 0.50 and adjusted accordingly the Rf of the bands of the other species. An unknown population should always be compared on the same slab with a population of a known species and the Rfs should be transformed accordingly. Only major bands of esterase activity appear to have taxonomic value. Many minor bands are also present in most species, but there is extensive variation among populations of the same species and the detection of these bands is not easy and often uncertain.

VI. Ecology

Most of the ecological information cannot be used directly in species identifications, but general differences can be helpful in eliminating certain species from consideration. Survival is influenced primarily by temperature, moisture, and suitability of available hosts. Generally climatic conditions that are favorable for a given host are equally favorable for the nematode. The ability of the four common species to attack so many different crop plants make these species widespread. Differences in host suitability can be useful taxonomically as previously discussed. Species not included in this guide are more host specific and thus limited in agricultural importance and in geographic distribution.

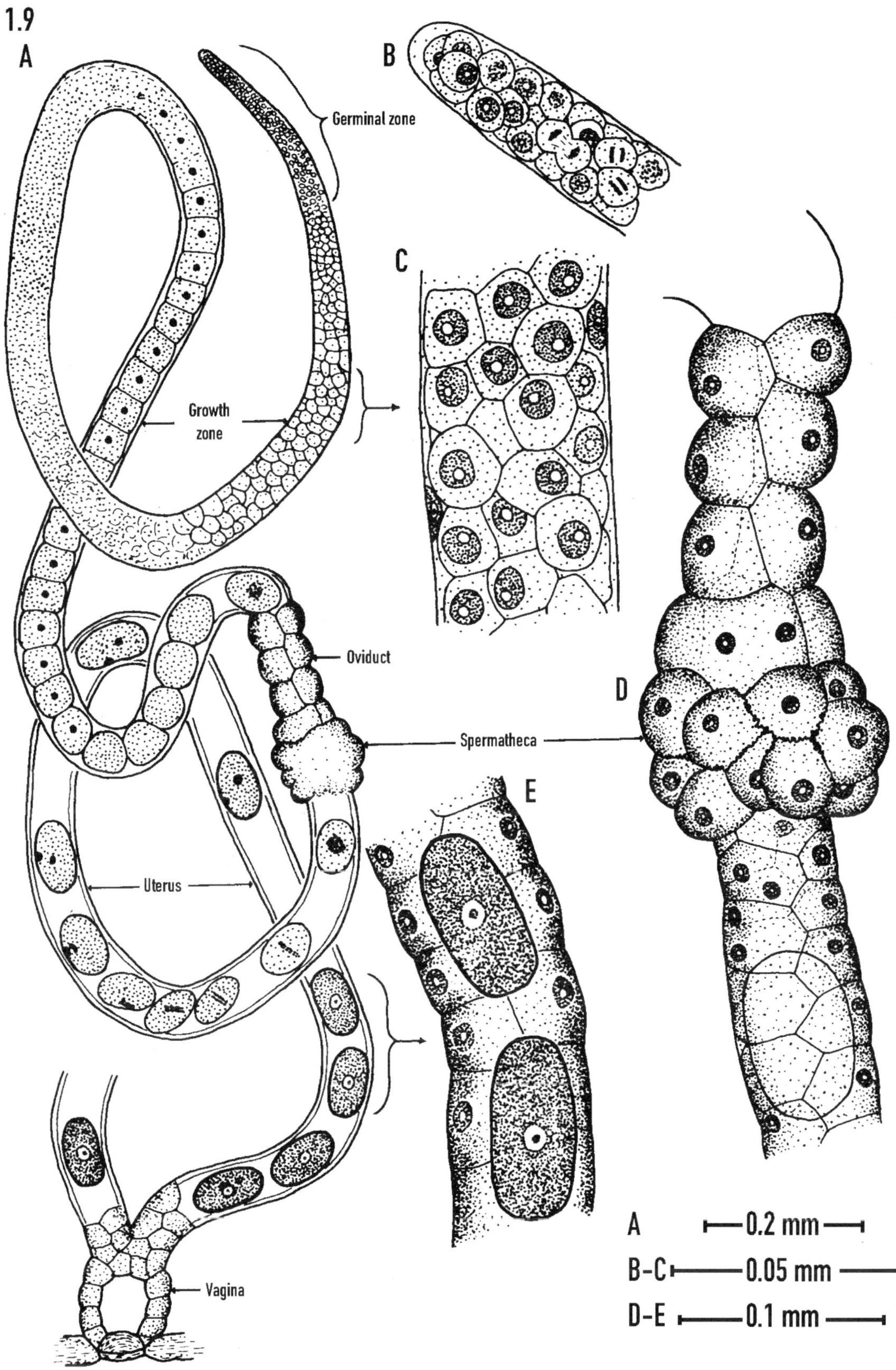

Fig. 1.9. A. Female reproductive system of *Meloidogyne javanica* showing the germinal zone, growth zone, oviduct, spermatotheca, uterus, and vagina. The reproductive systems of the other major species of *Meloidogyne* are identical to that of *M. javanica*. B, C, D, and E are enlarged drawings of the corresponding regions of the reproductive system. The process of oogenesis as illustrated in "A" involves a single mitotic division and is typical of all mitotically parthenogenetic forms of *Meloidogyne*. This pattern is modified in *M. incognita* and in *M. hapla* race A - See text. (After Triantaphyllou: Nematologica 7:105-113, 1962)

Major Bands of Esterases in Females

1.10

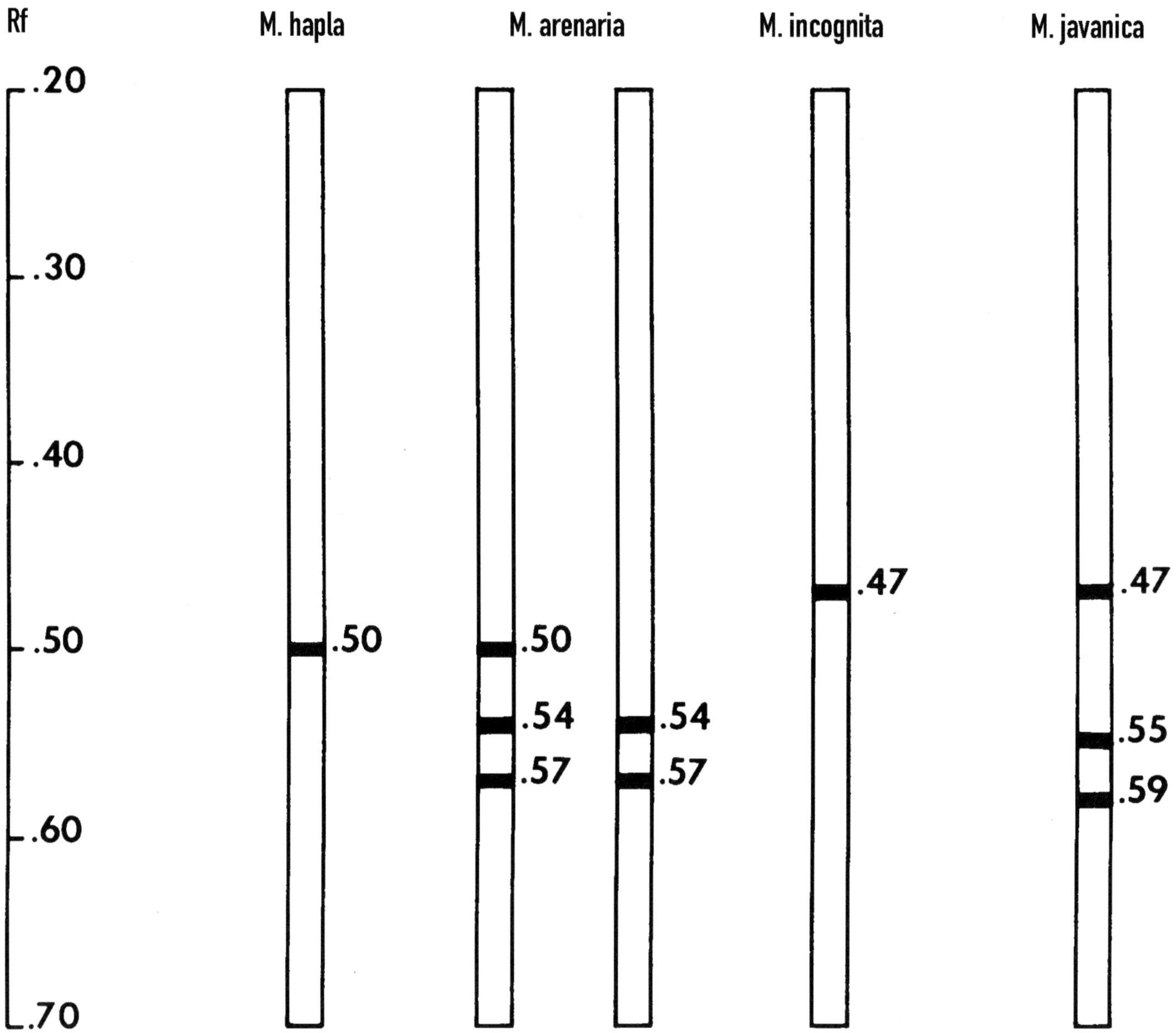

1

Fig. 1.10. Major bands of esterases in females of four species of *Meloidogyne* as revealed by acrylamide gel electrophoresis and a-Naphthyl acetate staining (Adapted from unpublished data by Janati, Berge, Dalmasso, and Triantaphyllou.)

Part 2. A Pictorial Key to the Four Common *Meloidogyne* Species

I. Introduction

This key is based on (A) morphology of perineal patterns, (B) head shape of males, (C) stylet morphology of males, and (D) differential host test. Although tentative identifications can be made from each character alone, one should consider as many characters as possible. With respect to the differential host test, the population in question must first be increased on a suitable host to obtain sufficient inoculum for inoculation of the various test plants. (See Taylor and Sasser, 1978, for guidelines for differential host test.)

Males should be obtained by incubating an infected, washed root system in a moist chamber at room temperature. Slide preparations of males and perineal patterns of adult females should be made to compare with the photographs shown in this guide. In case a mixed population of two species is present or the host tests are not typical, several perineal patterns as well as slides of males, if present, from each infected host should be prepared.

M. incognita

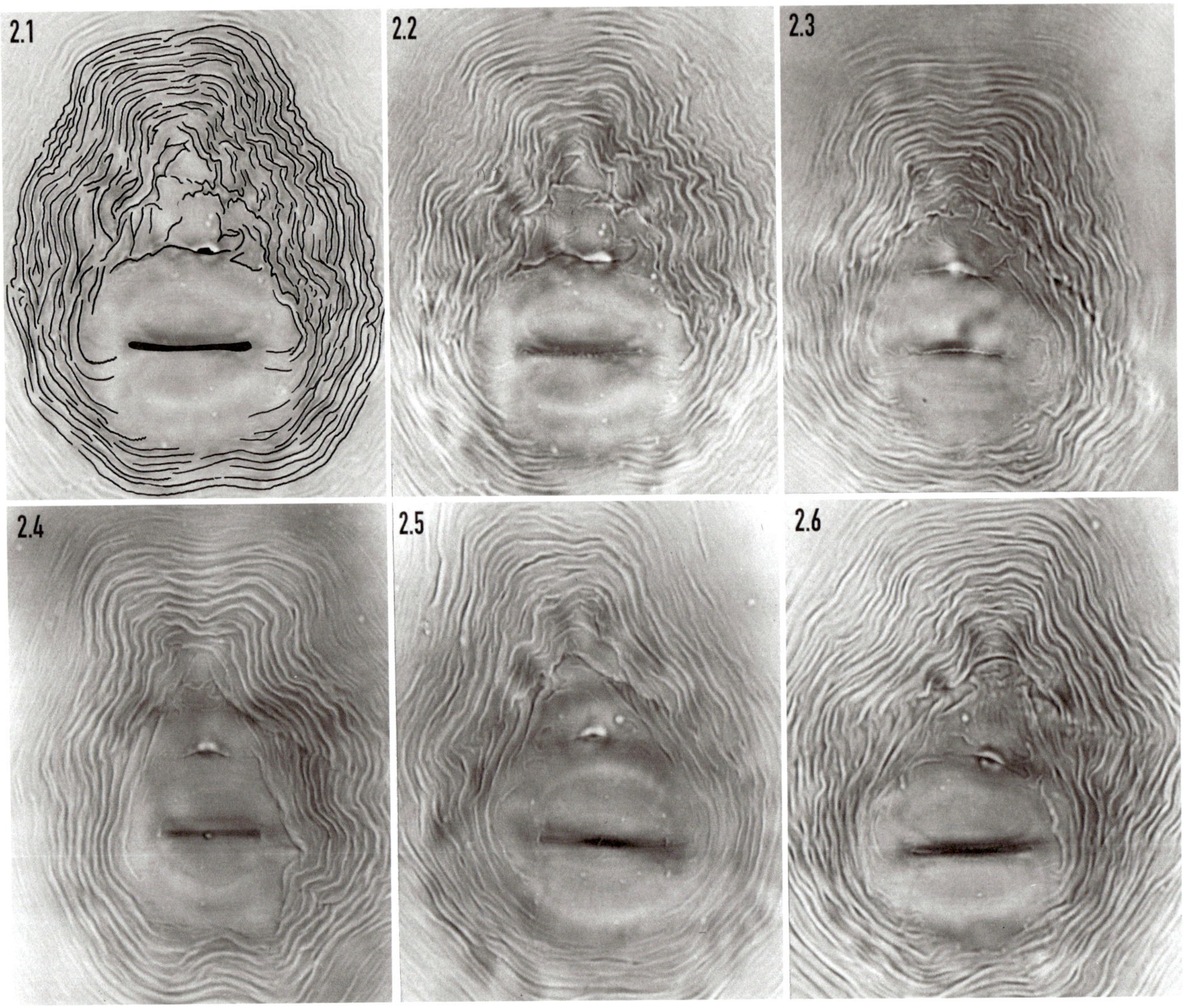

Figs. 2.1-2.6. *Meloidogyne incognita* perineal patterns. The high, squared-off dorsal arch is the key character for this species. Figures 2.1 and 2.2 are photographs of the same pattern, except in Fig. 2.1, the striae have been traced in ink

II. A Pictorial Key

I. (A) Perineal pattern with high dorsal arch, distinct lateral lines absent **(Figs. 2.1-2.6)**; (B) head of male with centrally concave labial disc raised above medial lips **(Fig. 2.25)**; (C) anterior portion of stylet of male paddle shaped, blunt; stylet knobs rounded to broadly elongate; distance from base of knobs to DGO short, 2-3 µm **(Fig. 2.25)**; (D) reproduction on pepper and watermelon but not peanut ....*M. incognita*

1. no reproduction on cotton or resistant tobacco .... race 1
2. reproduction on resistant tobacco but not cotton .... race 2
3. reproduction on cotton but not resistant tobacco .... race 3
4. reproduction on both cotton and resistant tobacco ....race 4

M. javanica

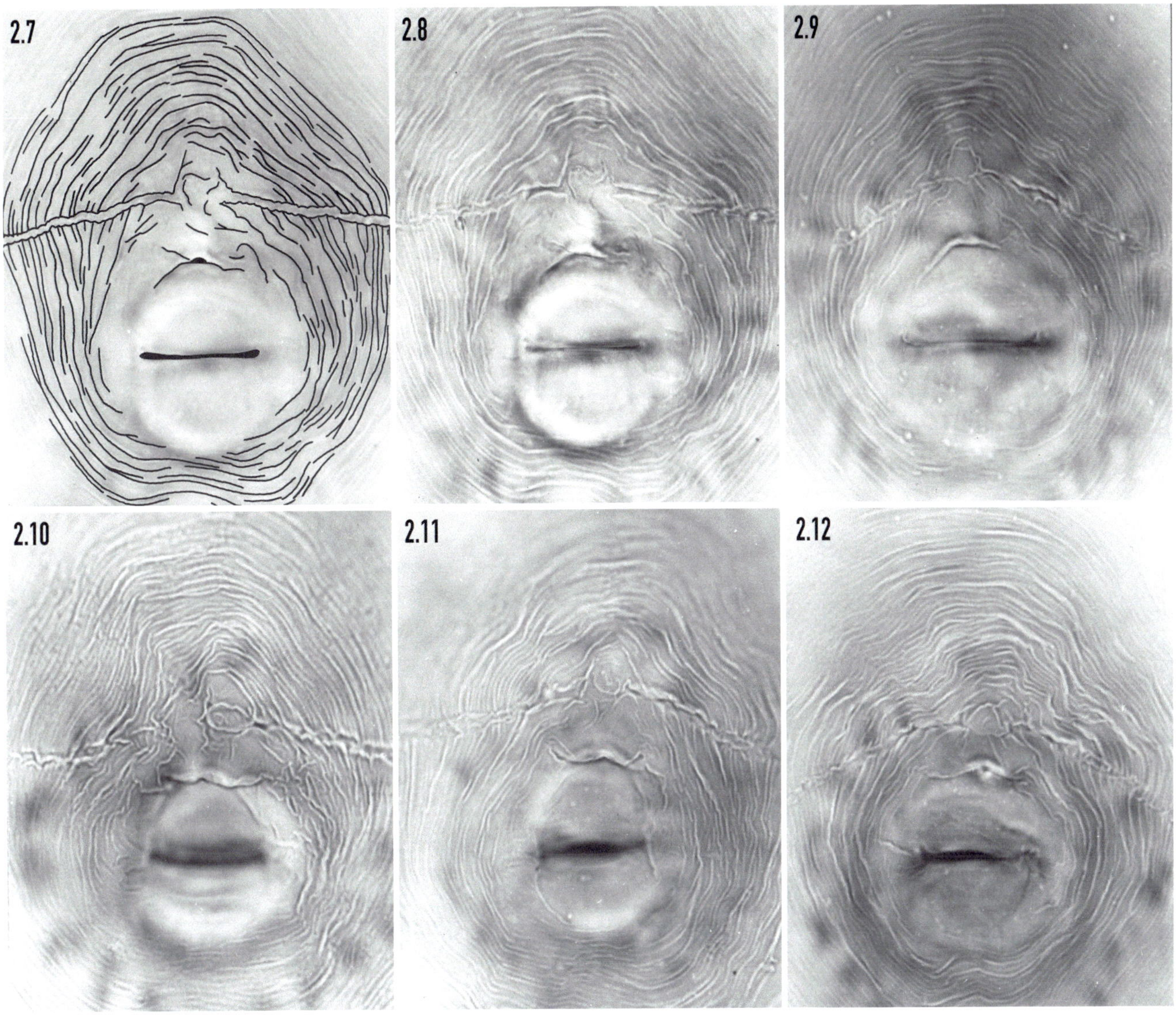

Figs, 2.7-2.12. ***Meloidogyne javanica*** **perineal patterns. The key character is the distinct lateral lines that separate the dorsal and ventral striae. Figure 2.7 is an ink tracing of the photograph shown in Fig. 2.8**

I.' (A) Perineal pattern with low to rounded arch, with or without distinct lateral lines (if lateral lines are present, arch may be high); (B) head of male with fused labial disc and medial lips (head cap) in the same contour; (C) stylet tip of male pointed; (D) reproduction on resistant tobacco but not cotton..........II

M. arenaria

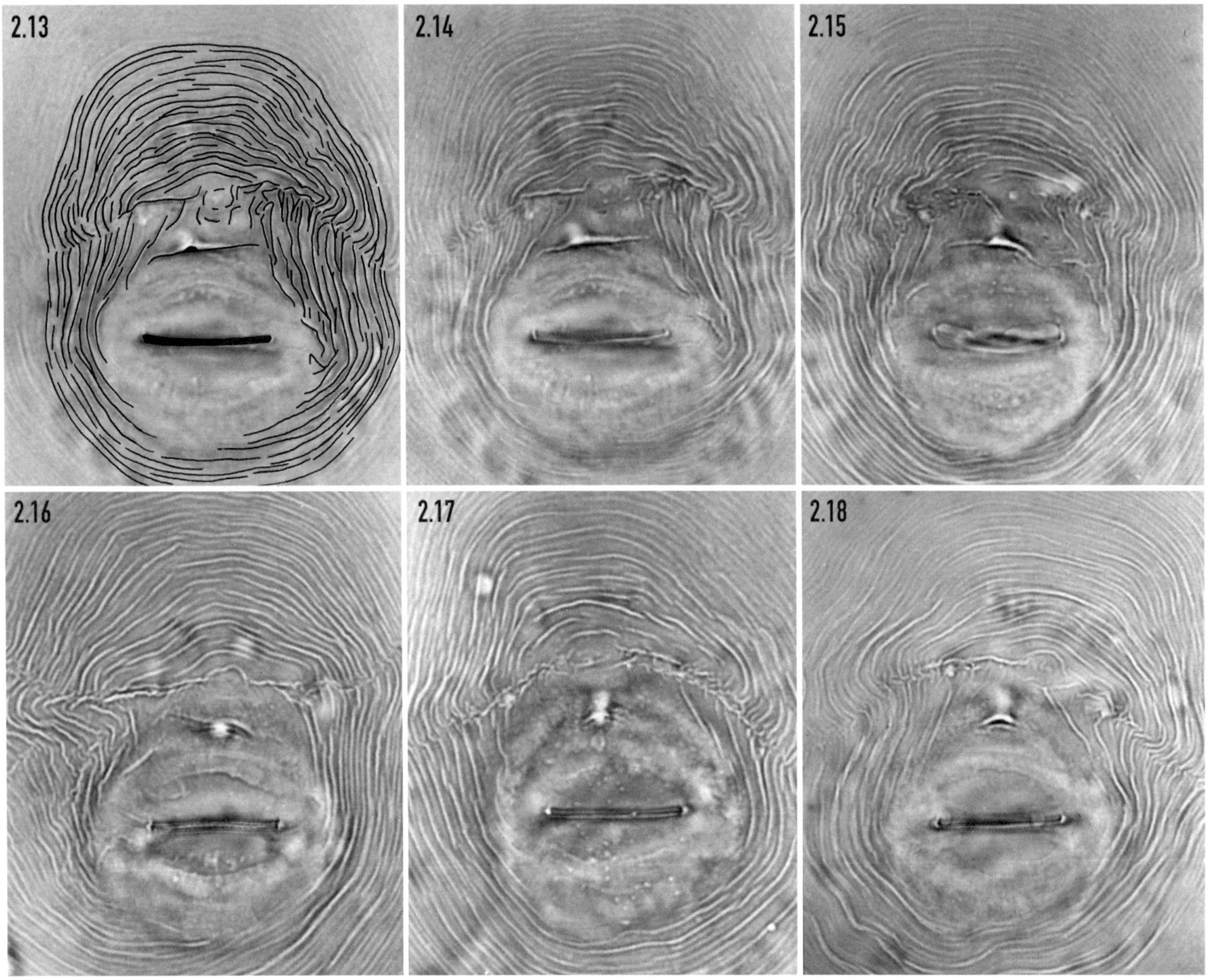

Figs. 2.13-2.18. ***Meloidogyne arenaria*** **perineal patterns. A dorsal arch with shoulders, formed by a slight indentation of the dorsal striae near the lateral lines, and striae that are forked near the lateral lines are the most important features of** ***M. arenaria*** **patterns. Figures 2.13 and 2.14 are photographs of the same pattern, except the striae have been traced in ink in Fig. 2.13**

II. (A) Perineal pattern with distinct lateral lines **(Figs. 2.7-2.12)**; (B) head cap of male high, nearly as wide as head region **(Fig. 2.26)**; (C) stylet knobs of male low and wide; distance from base of knobs to DGO short, 2- 3 m **(Fig. 2.26)**; (D) reproduction on watermelon but not pepper, cotton, or peanut ....*M. javanica*

M. hapla

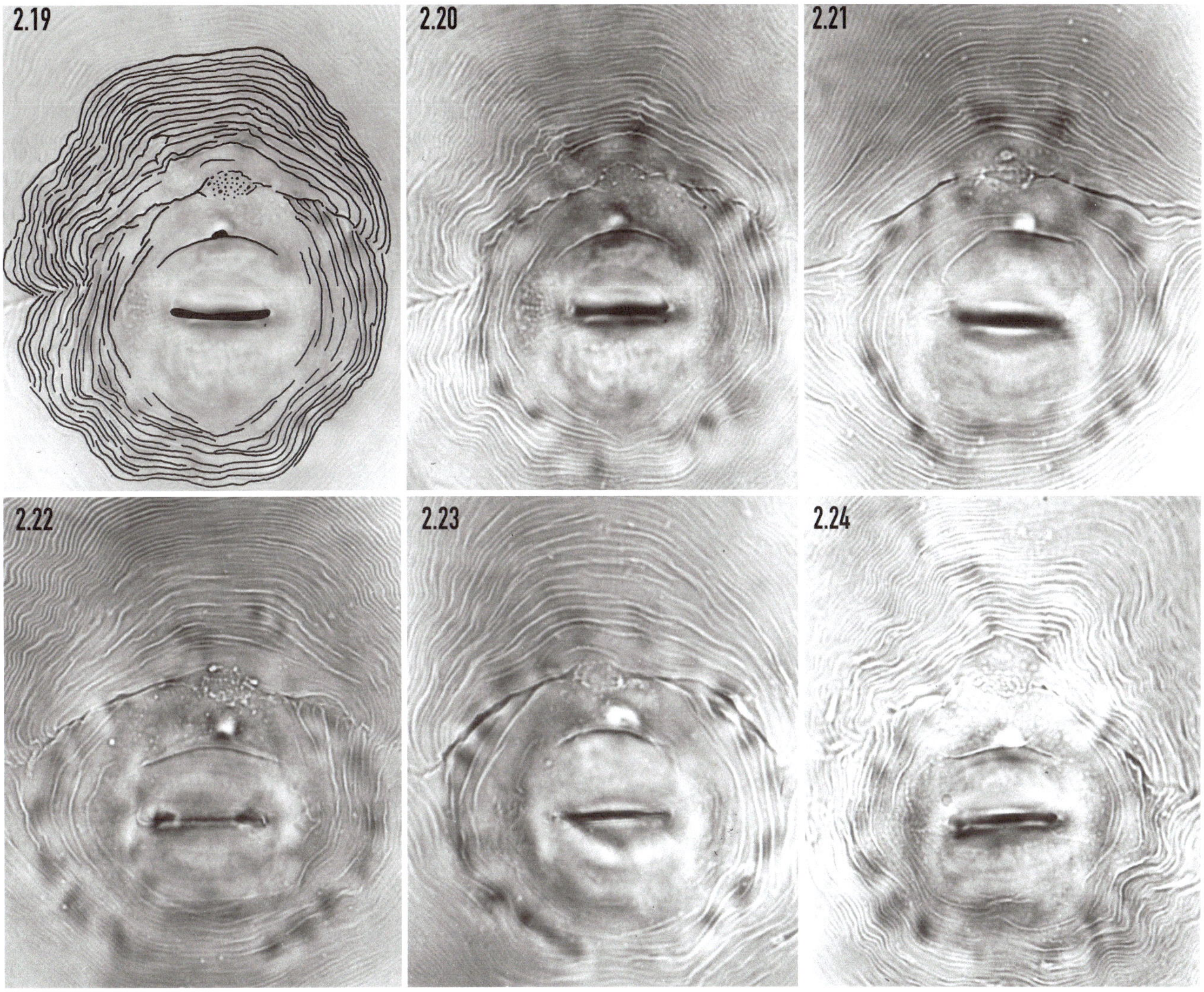

Figs. 2.19-2.24. ***Meloidogyne hapla*** **perineal patterns. The overall shape, a rounded hexagon to a flattened oval, and punctations in the tail terminal area are the key characters for this species (fixation may affect the appearance of punctations). Figure 2.19 is an ink tracing of the photograph shown in Fig. 2.20**

II.' (A) Perineal pattern without distinct lateral lines, arch rounded to flattened; (B) head cap of male low, sloping posteriorly or head cap high and not as wide as head region; (C) stylet knobs of male not low and wide; distance from base of knobs to DGO long, 4-7 µm; (D) reproduction on resistant tobacco and pepper......III

M. incognita

2.25

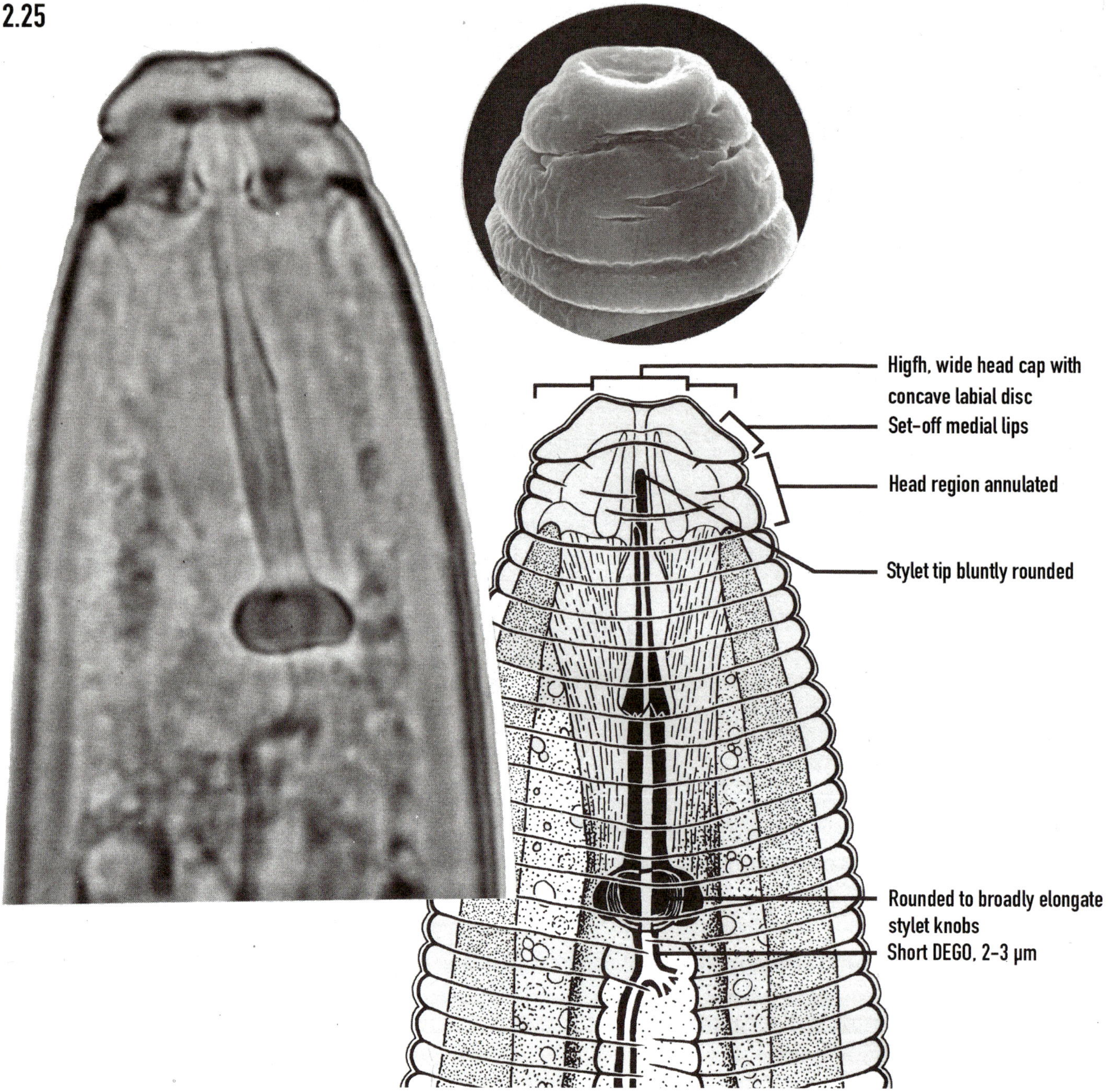

Fig. 2.25. LM and SEM photographs and line drawing of the head of a male of *Meloidogyne incognita*

III. (A) Dorsal and ventral striae forked meeting at an angle in lateral field, dorsal striae slightly indented for ming shoulder on arch **(Figs. 2.13-2.18)**; (B) head cap low, sloping posteriorly, nearly as wide as head region **(Fig. 2.27)**; (C) stylet knobs gradually merging with the shaft; distance from base of knobs to DGO long, 4-7 µm **(Fig. 2.27)**; (D) reproduction on resistant tobacco, pepper, and watermelon but not cotton .... *M. arenaria*

1. reproduction on peanut .... race 1
2. no reproduction on peanut .... race 2

M. javanica

2.26

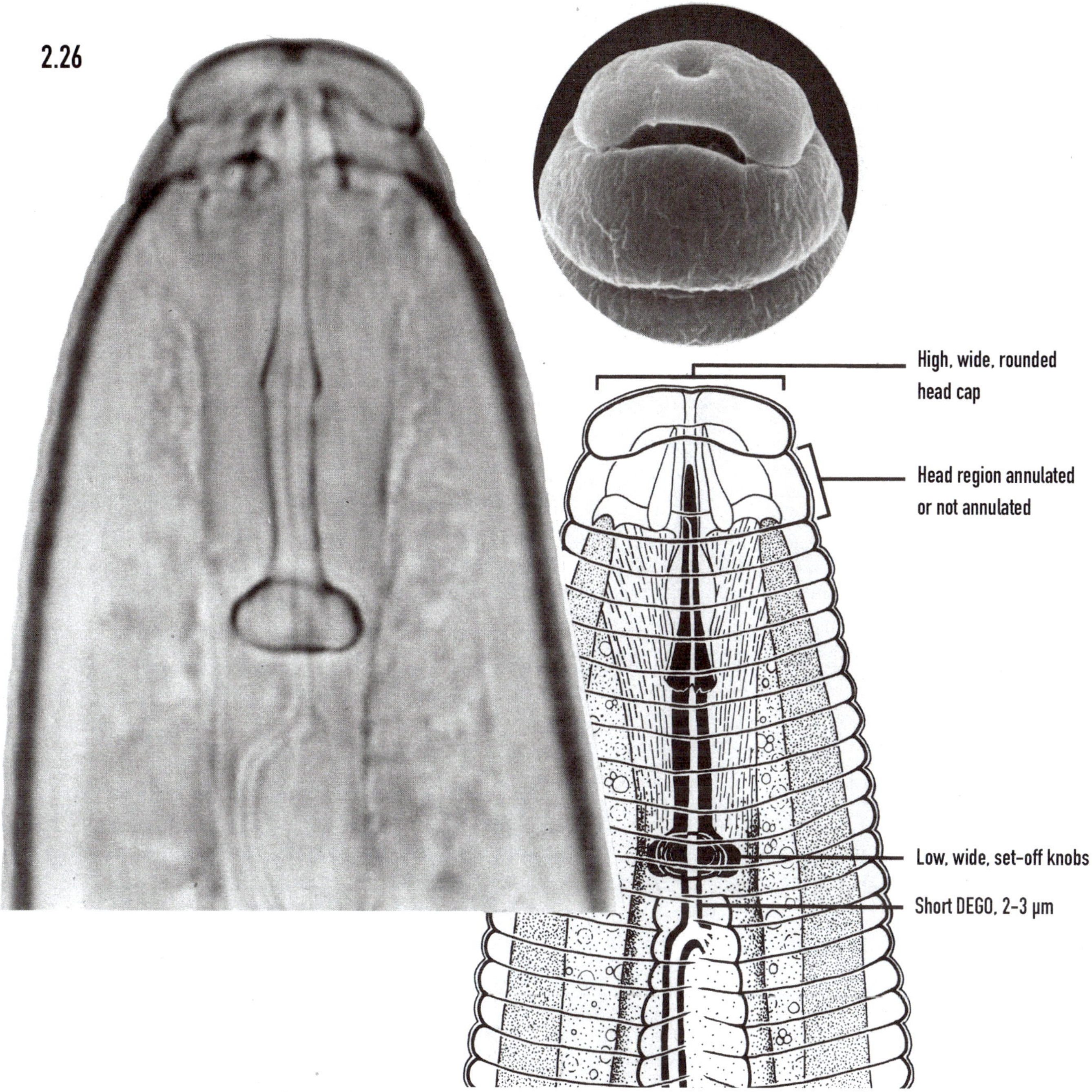

Fig. 2.26. LM and SEM photographs and line drawing of the head of a male of *Meloidogyne javanica*

III.' (A) Perineal pattern without shoulder on arch; (B) head cap of males high, not as wide as head region, not sloping posteriorly; (C) stylet knobs rounded, set off from shaft; (D) no reproduction on cotton or watermelon.........IV

2.27

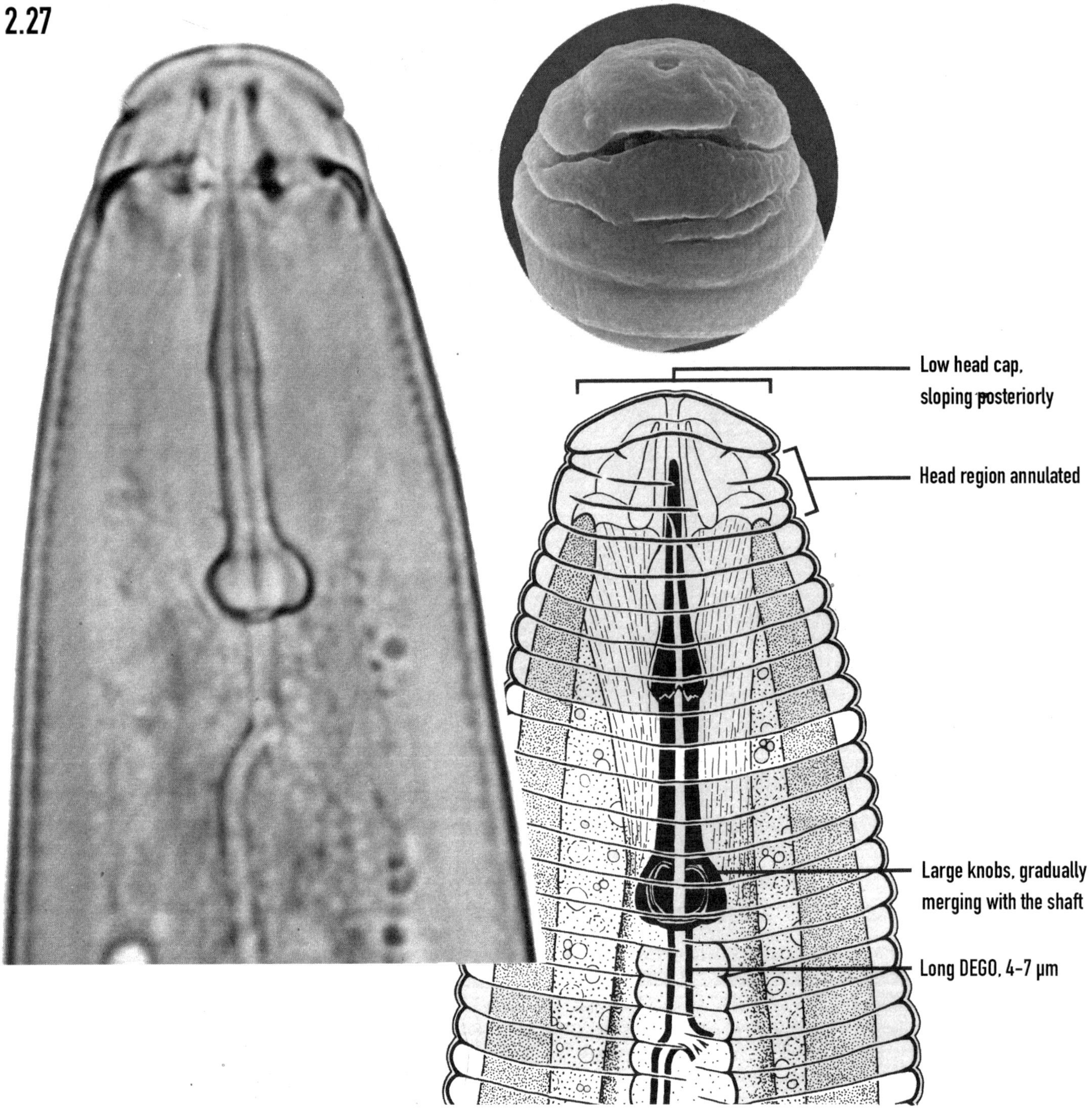

Fig. 2.27. LM and SEM photographs and line drawing of the head of a male of *Meloidogyne arenaria*

IV. (A) Perineal pattern nearly rounded hexagon to flattened oval, often with punctations in tail terminal area **(Figs. 2.19-2.24)**; (B) Usually head region of males set off from body annules, head cap not as wide as head region **(Fig. 2.28);** (C) stylet narrow, short; stylet knobs rounded, set off from shaft; distance from base of knobs to DGO long, 4-6 µm **(Fig. 2.28)**; (D) reproduction on resistant tobacco, pepper, and peanut but not watermelon or cotton.. *M. hapla*

M. hapla

2.28

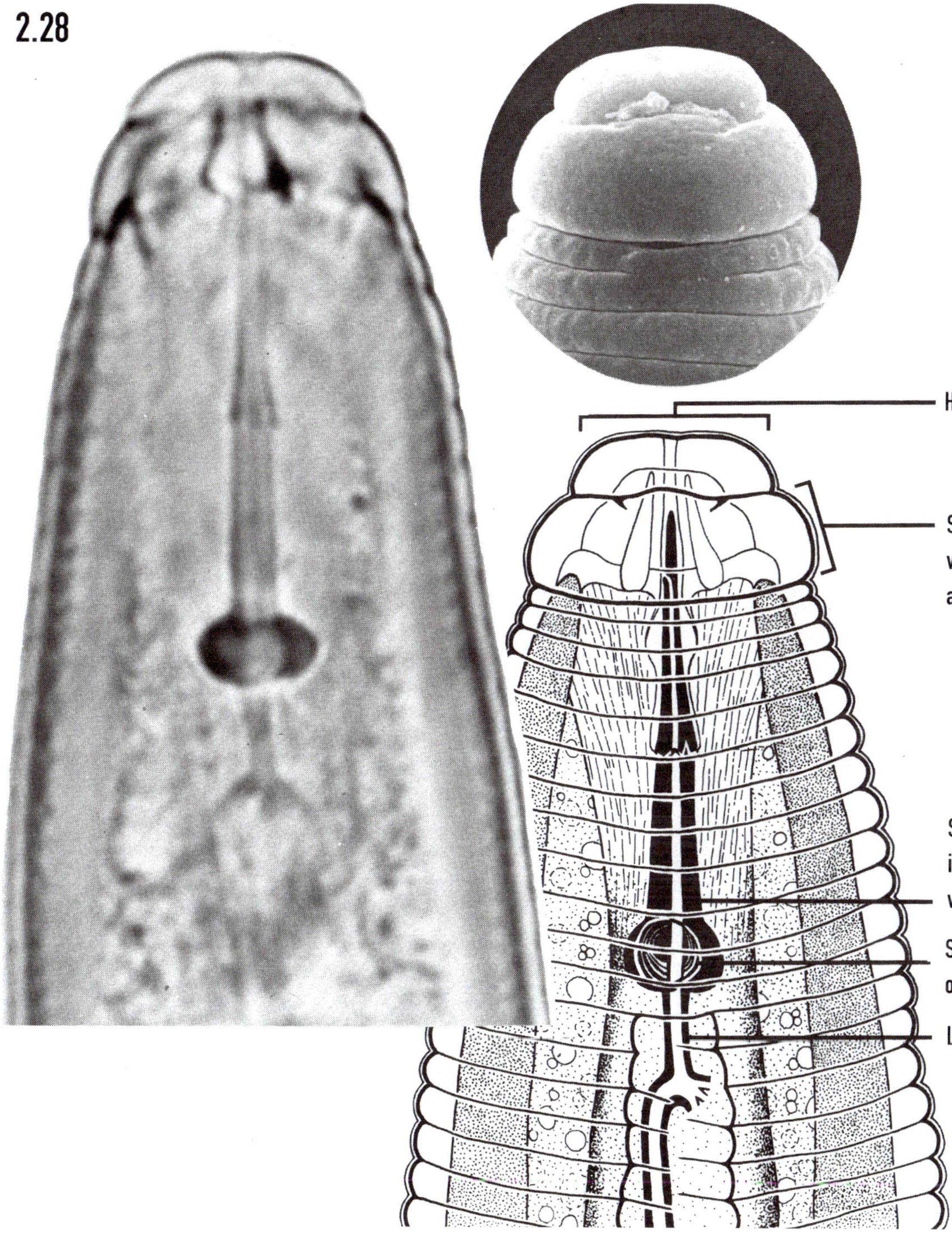

Fig. 2.28. LM and SEM photographs and line drawing of the head of a male of *Meloidogyne hapla*

IV.' Perineal pattern, male head shape, male stylet morphology, or host response different from above probably means the population does not belong to one of the four common species.

Part 3. A More Complete Characterization of the Four Most Common *Meloidogyne* Species

I. *Meloidogyne incognita*

A. Morphology.

All photographs are of IMP population 68 from North Carolina, host race 1 with 41- 43 chromosomes.

1. Females

a. Perineal patterns - Patterns of *M. incognita* **(Figs. 3.1-3.4)** have a distinct high dorsal arch composed of smooth to wavy striae. Some striae fork near the lateral lines but distinct lateral incisures are not present. Often there are striae that bend toward the vulva.

b. Stylets - In *M. incognita*, **(Figs. 3.5-3.7)** the stylet cone is distinctly curved dorsally. The anterior portion of the cone is cylindrical and the posterior half is conical. The shaft is slightly wider posteriorly. The stylet knobs are broadly elongate, set off from the shaft, and anteriorly indented, so much in some specimens that each knob appears as two.

c. Head morphology (SEM) - The labial disc and medial lips of *M. incognita* **(Figs. 3.8-3.9)** are dumb-bell-shaped (the medial lips are wider than the labial disc) in face view. Two bumps are present on the ventral side of the labial disc. The lateral lips are large and separated from the rounded medial lips and generally fuse with the head region for a short distance laterally. The head region is often marked by one broken annulation.

2. Males

a. Head morphology - The head shape of M. incognita males **(Figs. 3.10-3.13)** is very characteristic and not easily confused with any other species. The labial disc is large and round, centrally concave, and raised above the medial lips. The medial lips are as wide as the head region which is generally marked by 2 or 3 incomplete annulations.

b. Stylets - The stylet tip of *M. incognita* males **(Figs. 3.12-3.14)** is blunt and wider than the medial portion of the cone. A projection on the ventral side of the cone marks the opening of the stylet lumen which is located one fourth the distance of the cone length from the stylet tip. The shaft is generally cylindrical and often narrows near the knobs. The knobs are set off from the shaft, anteriorly indented, and broadly elongate to round.

3. Second-Stage Juveniles

Head morphology (SEM) - *M. incognita* second-stage juveniles **(Figs. 3.15-3.16)** have a dumbbell shaped labial disc and medial lips in face view. The labial disc is small and round, slightly raised above the medial lips. Lateral lips lie in contour with the head region which usually bears two to four incomplete annulations.

4. Useful Measurements (from three populations, 50 specimens each)

Second-stage juvenile total length, 346-463 (405) µm, range (mean); tail length, 42-62 (52) µm; head end to stylet base 14-16 (15) µm; female stylet length, 15-17 (16) µm; male stylet length 23-25 (24) µm.

B. Differential Host Test

According to the North Carolina Differential Host Test, *M. incognita* is made up of four host races. Populations of all four races reproduce on pepper, watermelon, and tomato; but they vary in their response to resistant tobacco and cotton. Race 1 populations do not reproduce on tobacco or cotton, race 2 populations reproduce on tobacco but not on cotton, race 3 populations do not reproduce on tobacco but do reproduce on cotton, and race 4 populations reproduce on tobacco and cotton. The reactions of populations of the four races of *M. incognita* are summarized in **Table 3.1.**

Table 3.1. Differential host test identification for races of *Meloidogyne incognita*.

Meloidogyne incognita race	Cotton Deltapine 16	Tobacco NC95
Race 1	–	–
Race 2	–	+
Race 3	+	–
Race 4	+	+

C. Symptomatology

In susceptible plants, *M. incognita* populations may produce galls that occur singly, but usually the galls coalesce to form large and sometimes massive galls (e.g. on cucurbits). Generally the galling is not considered to be useful in identification to species.

M. incognita

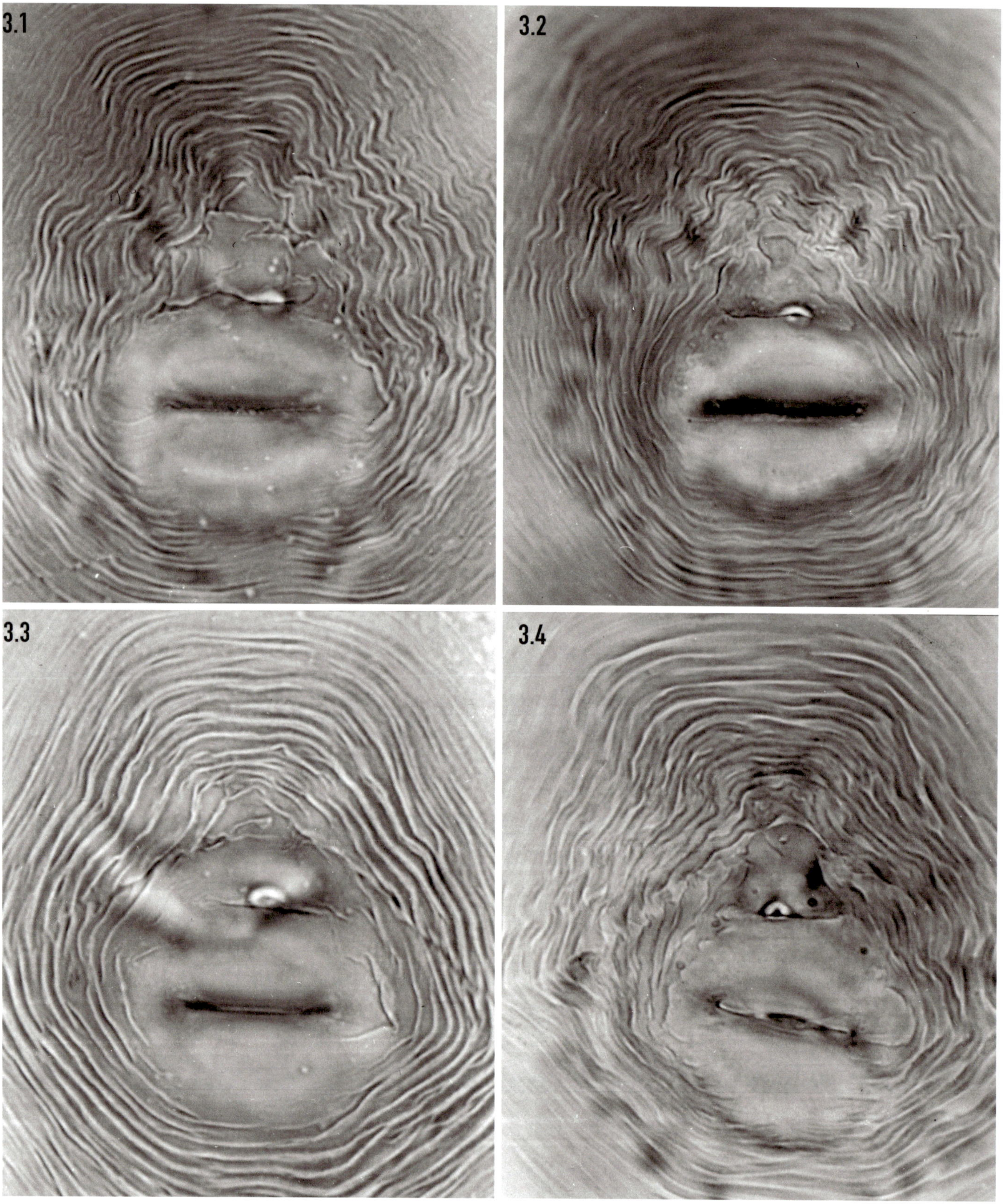

Figs. 3.1-3.4. Perineal patterns of *Meloidogyne incognita*

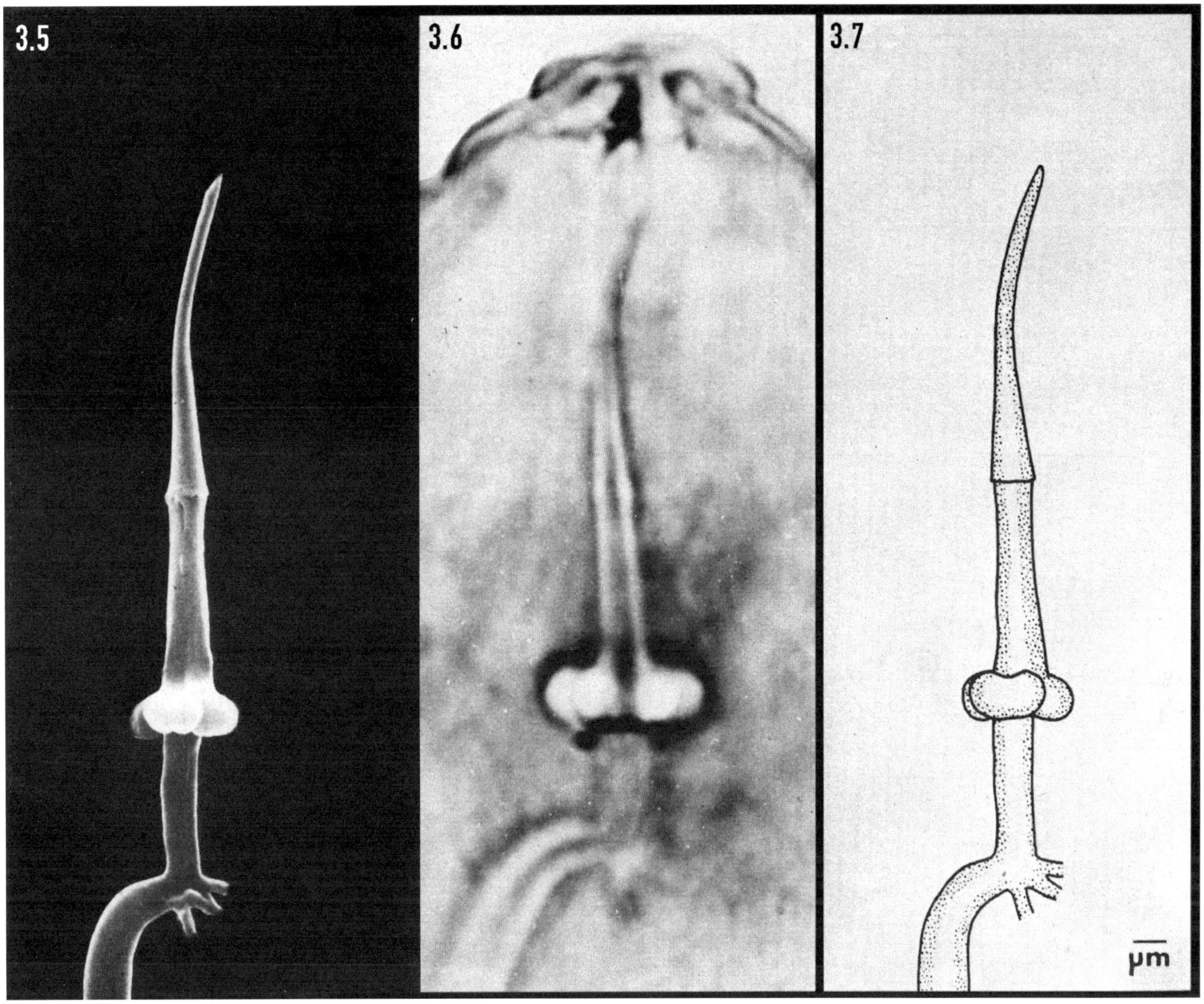

Figs. 3.5-3.7. SEM and LM photographs and line drawing of a stylet of a female of *Meloidogyne incognita*

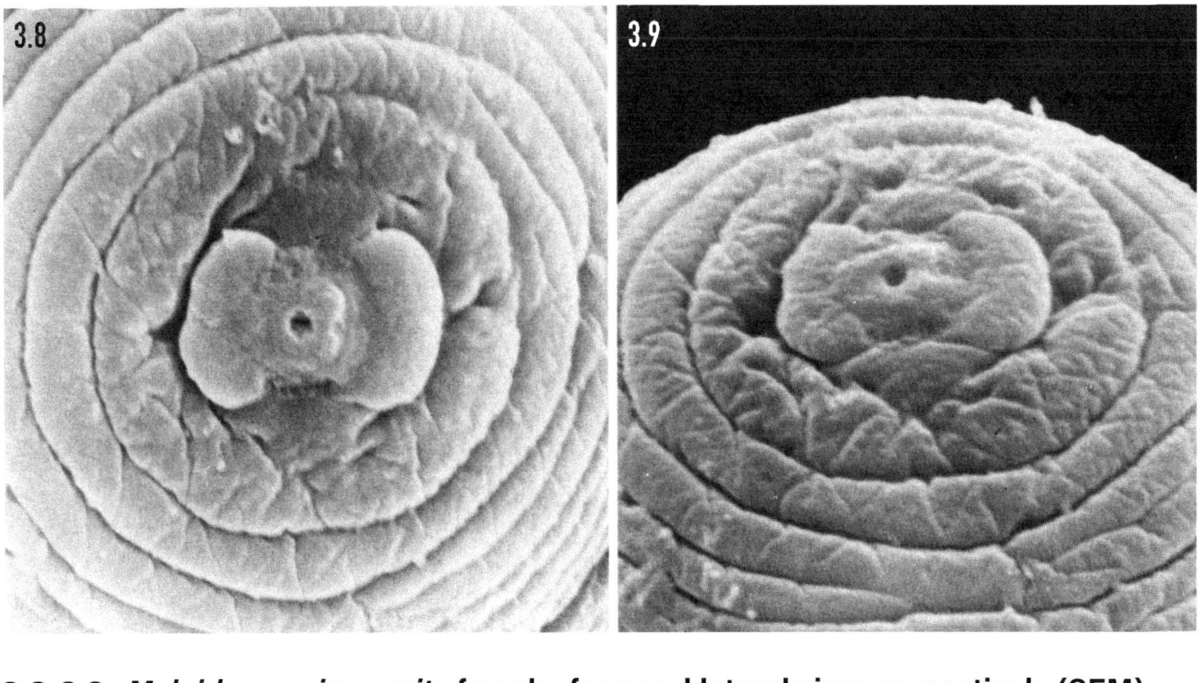

Figs. 3.8-3.9. *Meloidogyne incognita* **female, face and lateral view, respectively (SEM)**

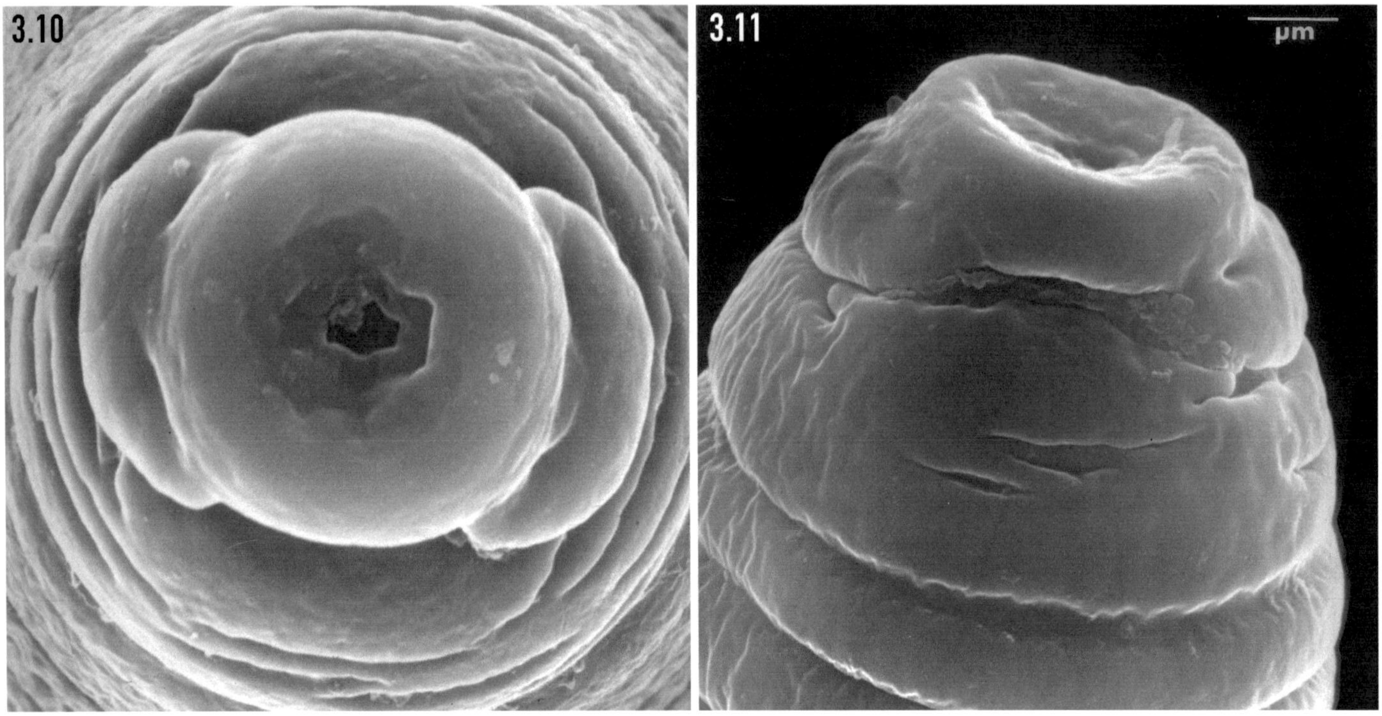

Figs. 3.10-3.11. *Meloidogyne incognita* male, face and lateral view, respectively (SEM)

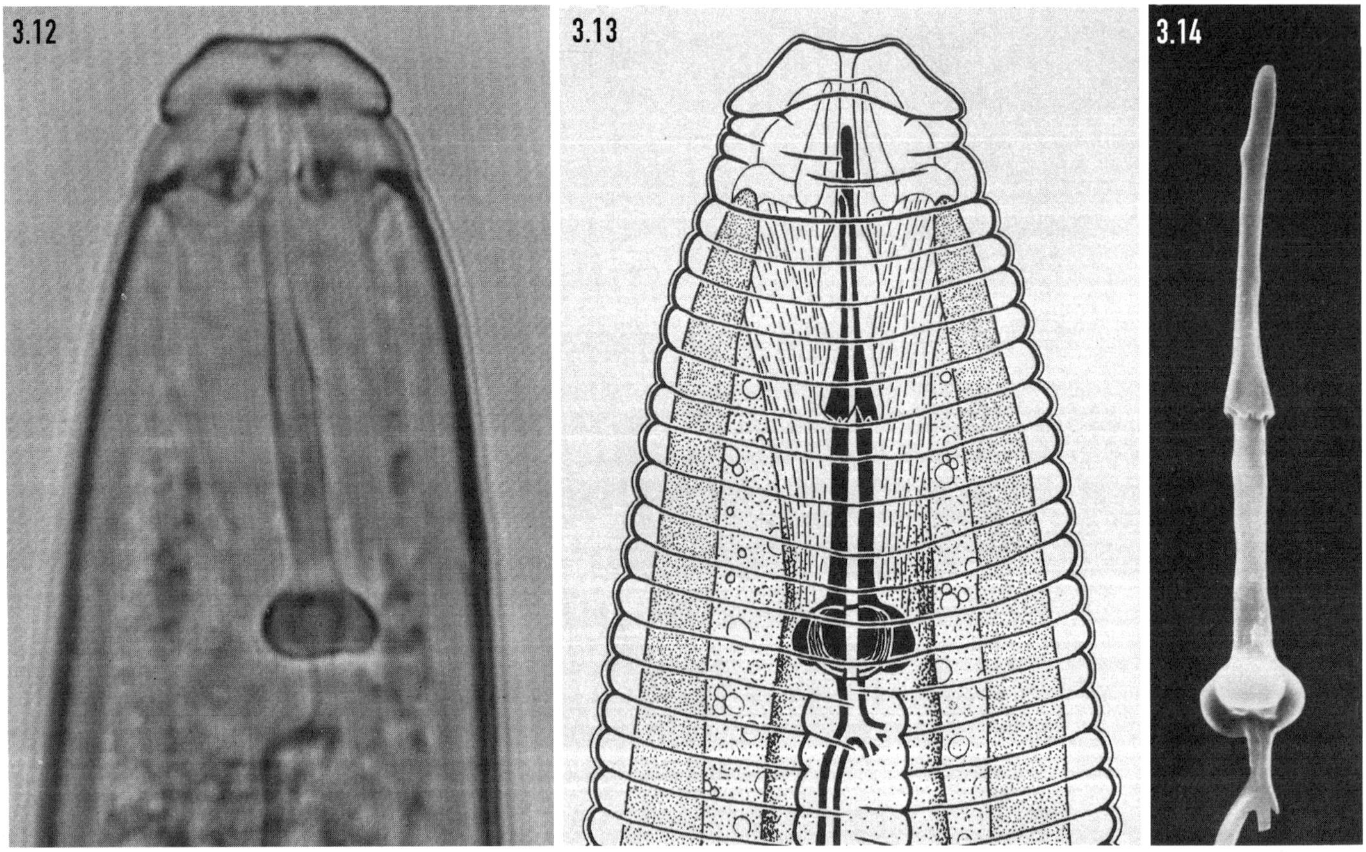

Figs. 3.12-3.13. LM photograph and line drawing of the head and stylet of a *Meloidogyne incognita* male. **3.14.** SEM photograph of an excised stylet of a male of *M. incognita*

D. Cytogenetics

Meloidogyne incognita populations reproduce exclusively by mitotic parthenogenesis. There are two chromosomal forms within this species. One form has 2n=32-36 chromosomes and is considered to be diploid; the other form has 2n=40-46 chromosomes and probably represents a triploid. The triploid form is by far the most common and widely distributed around the world. All populations of *M. incognita* have a unique cytological feature that separates them from populations of all other species of *Meloidogyne*. The oocytes of *M. incognita* are at prophase as they pass through the spermatotheca and remain in this stage until they have migrated to the posterior part of the uterus, when they suddenly advance to metaphase. During all this prolonged period of prophase, the chromosomes are bunched close together and cannot be seen individually or counted **(Figs. 3.17-3.18)**. Oocytes of all other *Meloidogyne* species advance to metaphase as soon as they pass through the spermatotheca into the uterus **(Fig. 1.9)**. Furthermore, the chromosomes are spread out in a large area, are discrete, and can be counted.

E. Biochemistry

A single, major band of esterase activity at Rf=0.47 is characteristic of *M. incognita* **(Fig. 1.10)**. No deviations from this basic pattern have been detected among 20 populations of diverse origin studied thus far. Variation exists in minor bands which probably are not characteristic of the species.

F. Ecology

Meloidogyne incognita constitutes about 52% of the *Meloidogyne* species collected through the IMP. *M. incognita* occurs over a wider geographical area than the other species (approximately 40°N latitude to 33°S latitude) and has a very extensive host range. This species occurs where the average annual temperature range is between 18° to 30° C, with the greatest number of populations coming from areas where the range is 24° to 27°C (47%). Optimum warm month temperature is approximately 27°C. This species frequently is found coexisting with *M. javanica*. From a total of 423 *M. incognita* populations studied, 72% were host race 1; 15% race 2; 11% race 3; and 2% race 4. Thus 87% of *M. incognita* populations do not reproduce on cotton, 83% do not reproduce on resistant tobacco and none of the four races reproduces on peanut.

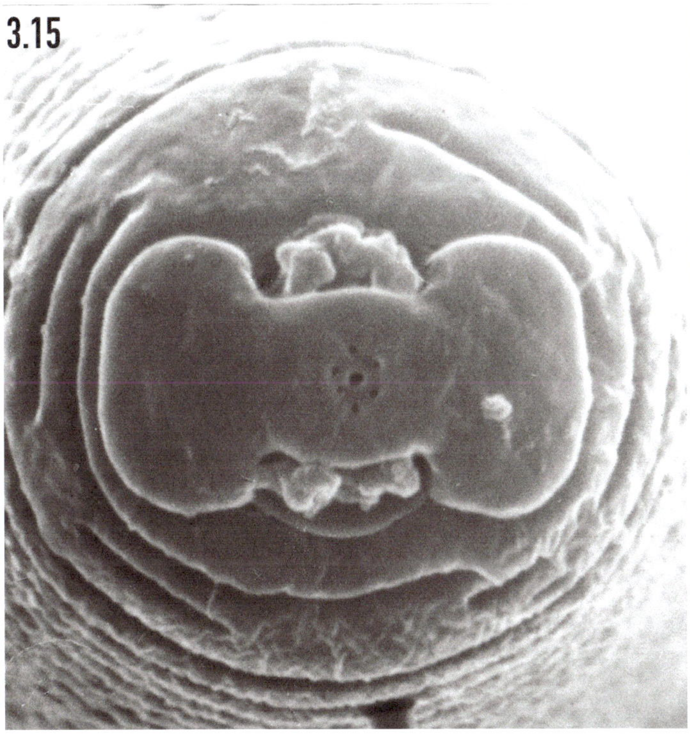

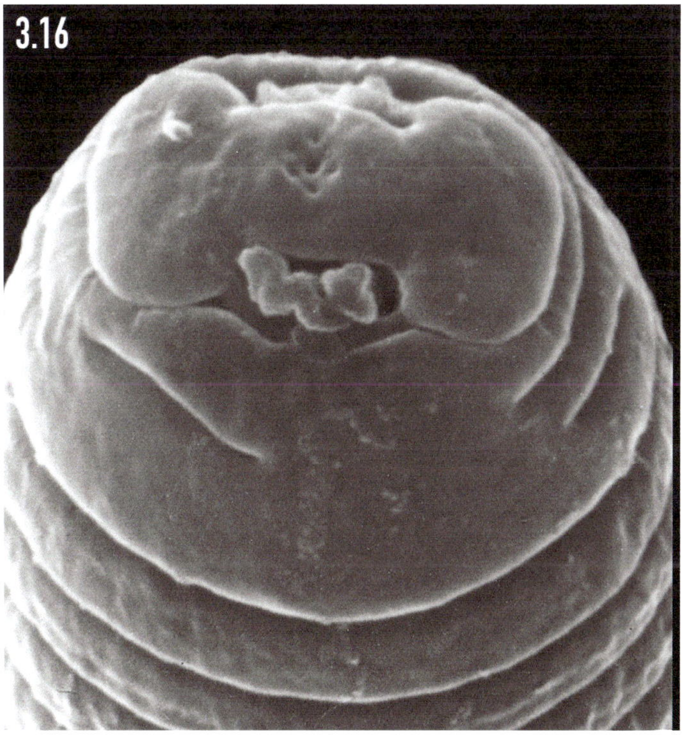

Figs. 3.15-3.16. Second-stage juvenile of *Meloidogyne incognita*, face and lateral view, respectively (SEM)

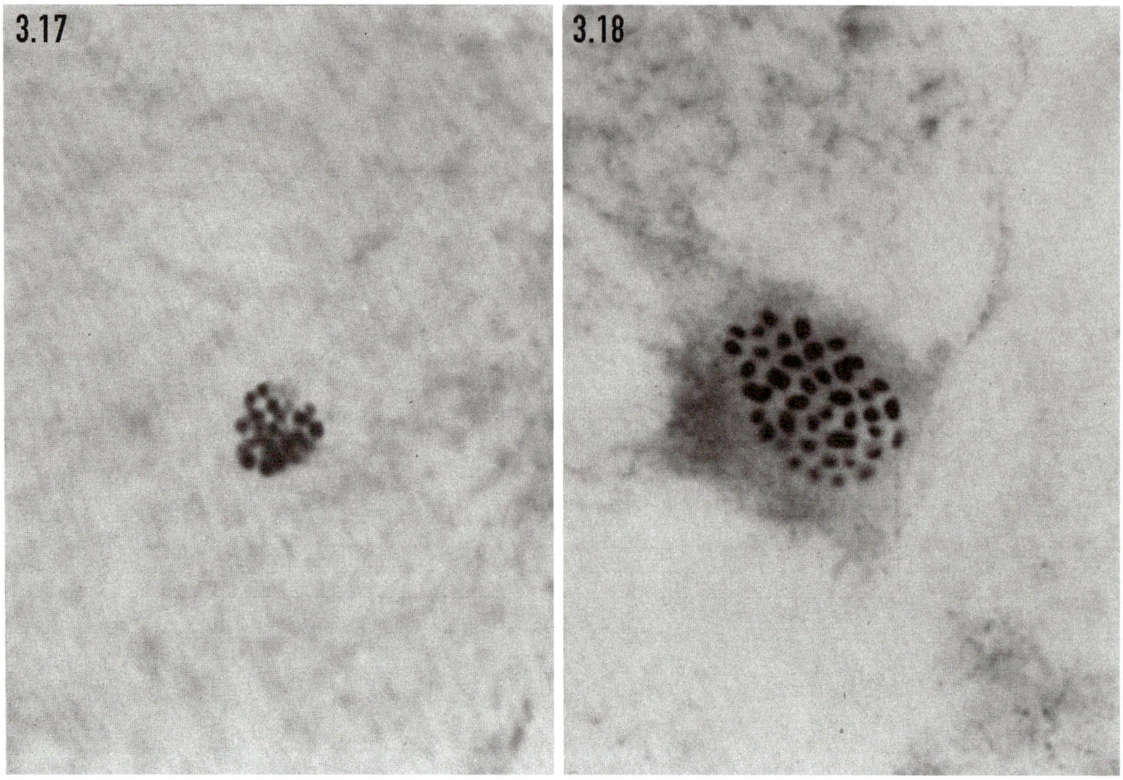

Figs. 3.17-3.18. Prophase and metaphase chromosomes in maturing oocytes of *M. incognita.* Most oocytes present in the uteri are at prophase with the chromosomes grouped together (Fig. 3.17). Only one or two oocytes in the posterior part of each uterus are at metaphase (Fig. 3.18) and show discrete chromosomes

II. *Meloidogyne javanica*

A. Morphology

All photographs are from IMP population 76 from Georgia with 44 chromosomes.

1. Females

a. **Perineal patterns** - *M. javanica* patterns, **(Figs. 3.19-3.22)** have a rounded to flattened dorsal arch. The characteristic feature of this pattern is distinct lateral incisures that separate the pattern into dorsal and ventral regions. No or few striae cross the lateral incisures and some striae bend toward the vulva.

b. **Stylets** - Stylets of *M. javanica* females **(Figs. 3.23-3.25)** are similar to those of *M. incognita*, except ▪ that the cone is not distinctly curved dorsally and gradually increases in width posteriorly. The shaft widens only slightly posteriorly, and the knobs are short and wide, often anteriorly indented.

c. **Head morphology** (SEM) - In *M. javanica*, **(Figs. 3.26-3.27)** the labial disc and medial lips are dumbbell shaped. The labial disc h s two prominent bumps ventrally. Usually the medial lips are indented which suggests a division of the lips into pairs of medial lips. The lateral lips are large, elongate, and set off from the medial lips and head region. Often the head region is marked by one incomplete annulation.

2. Males

a. **Head morphology** - In *M. javanica*, **(Figs. 3.28-3.31)** the large smooth labial disc and medial lips are fused. The head cap is high and almost as wide as the head region. In this particular population, the head region is not annulated, but some other populations of *M. javanica* have 2-3 head annules.

b. **Stylets** - The stylet cone of *M. javanica* males (**Figs. 3.30-3.32**) is narrow anteriorly, but very wide posteriorly. The shaft is cylindrical and often narrows near the junction with the stylet knobs. The stylet knobs are low and wide and set off from the shaft.

3. Second-Stage Juveniles

Head morphology (SEM). The labial disc and medial lips of *M. javanica* second-stage juveniles (**Figs. 3.33-3.34**) are bow-tie shaped. The lateral lips are triangular and lie below the contour of the labial disc and medial lips. Occasionally, the head region may have a short annulation, but generally it is smooth.

4. Additional Morphological Features

Some populations of *M. javanica* produce male intersexes that are characteristic for the species. These intersexes show different degrees of female secondary sex characters varying from a small ventral protuberance anterior to the spicules to a large protuberance marked by a rudimentary vulva. Some populations produce almost exclusively intersexes while others produce only normal males.

5. Useful measurements (from three populations, 30 specimens each)

Second-stage juvenile total length, 402-560 (488) µm, range (mean); tail length, 51-63 (56) µm; head end to stylet base 14-16 (15) µm; female styiet length, 14-18 (16) µm; male stylet length, 18-22 {20) µm.

B. Differential Host Test

North Carolina Differential Host Test results show that populations of *M. javanica* reproduce on root-knot resistant tobacco, watermelon, and tomato. Most *M. javanica* populations do not reproduce on cotton, pepper, or peanut. A few of the populations can reproduce on pepper, and likewise even fewer can reproduce on peanut.

C. Symptomatology

Galls produced by *M. javanica* populations are similar to those of *M. incognita* and are not considered to be diagnostic for the species.

D. Cytogenetics

Populations of *M. javanica* reproduce exclusively by mitotic parthenogenesis. The chromosome number varies from 2n=43 to 48. All populations belong to the same chromosomal form, which may represent a triploid. At metaphase of the single maturation division the chromosomes of *M. javanica* are univalents (dyads), spread in a lateral metaphase plate, and can be counted easier than those of any other species **(Fig. 3.35)**. Usually, two to four oocytes located in the uterus, close to the spermatotheca, are at metaphase and can be studied. All other oocytes in the uterus have advanced to anaphase and telophase and are of limited value for cytological study.

E. Biochemistry

Three major bands of esterase activity at Rf=0.47, 0.55 and 0.59 are typical of *M. javanica* **(Fig. 1.10)**. No variation in this pattern has been observed among 20 populations of different origin examined thus far.

F. Ecology

Meloidogyne javanica, with only one distinct host race, comprises about 31% of the *Meloidogyne* species collected through the IMP. Like *M. incognita*, this species has an extensive host range, but the latitude range is about 3 degrees less, from approximately 33° North to 33° South. In regions where the rainfall is more or less evenly distributed over the year, *M. javanica, M. incognita*, and occasionally *M. arenaria* are found in the same field populations. In regions with a distinct dry season and less than 5 mm of precipitation per month for three or more successive months, *M. javanica* may be the predominant species. *M. javanica* does not reproduce on strawberry, cotton, or peanut and rarely on pepper. Thus a root-knot nematode reproducing on these crops, or one found in the northern United States, Canada, United Kingdom or similar countries with similar temperatures, would probably not be *M. javanica*.

M. javanica

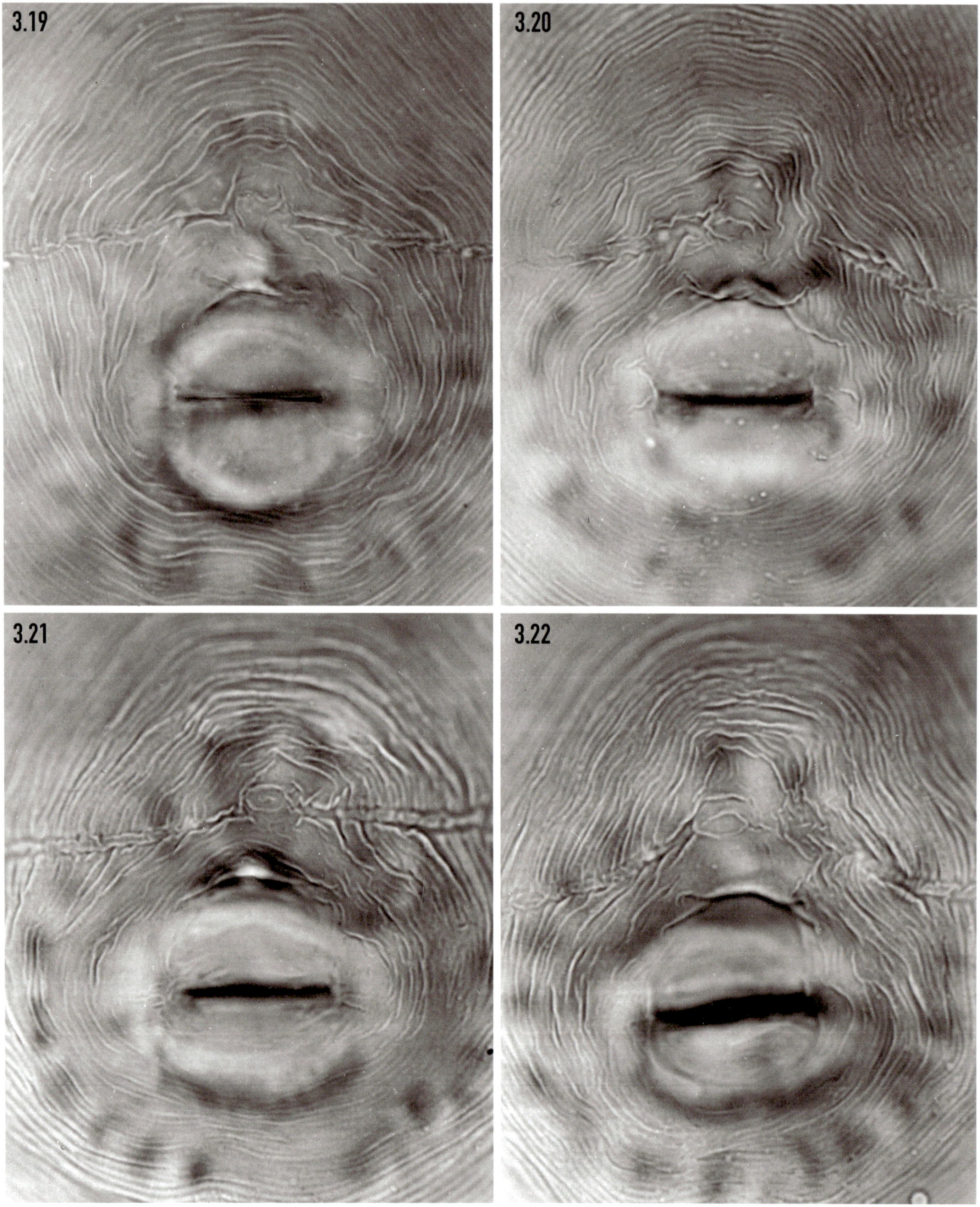

Figs. 3.19-3.22. Perineal patterns of *Meloidogyne javanica*

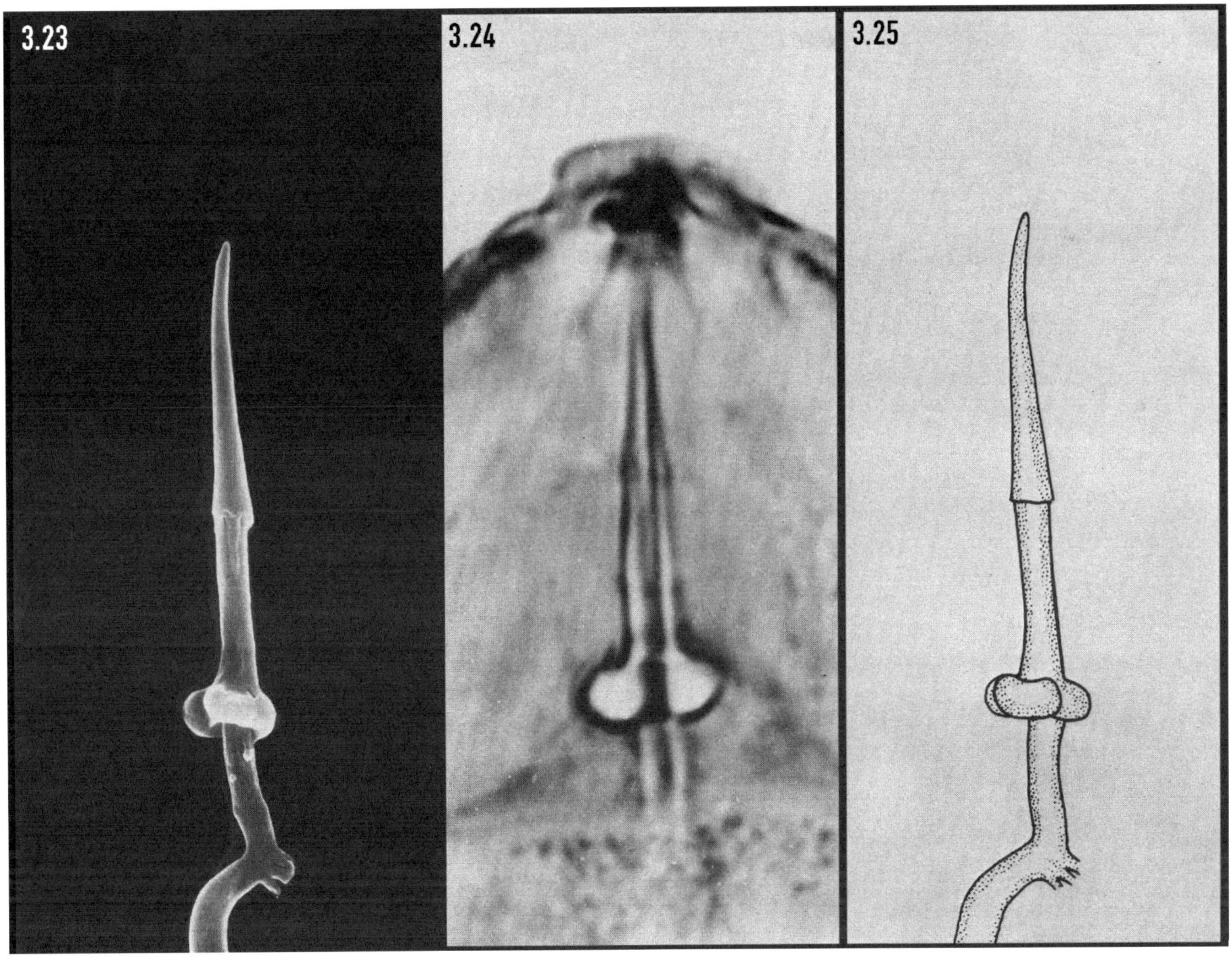

Figs. 3.23-3.25. SEM and LM photographs and line drawing of a stylet of a female of *Meloidogyne javanica*

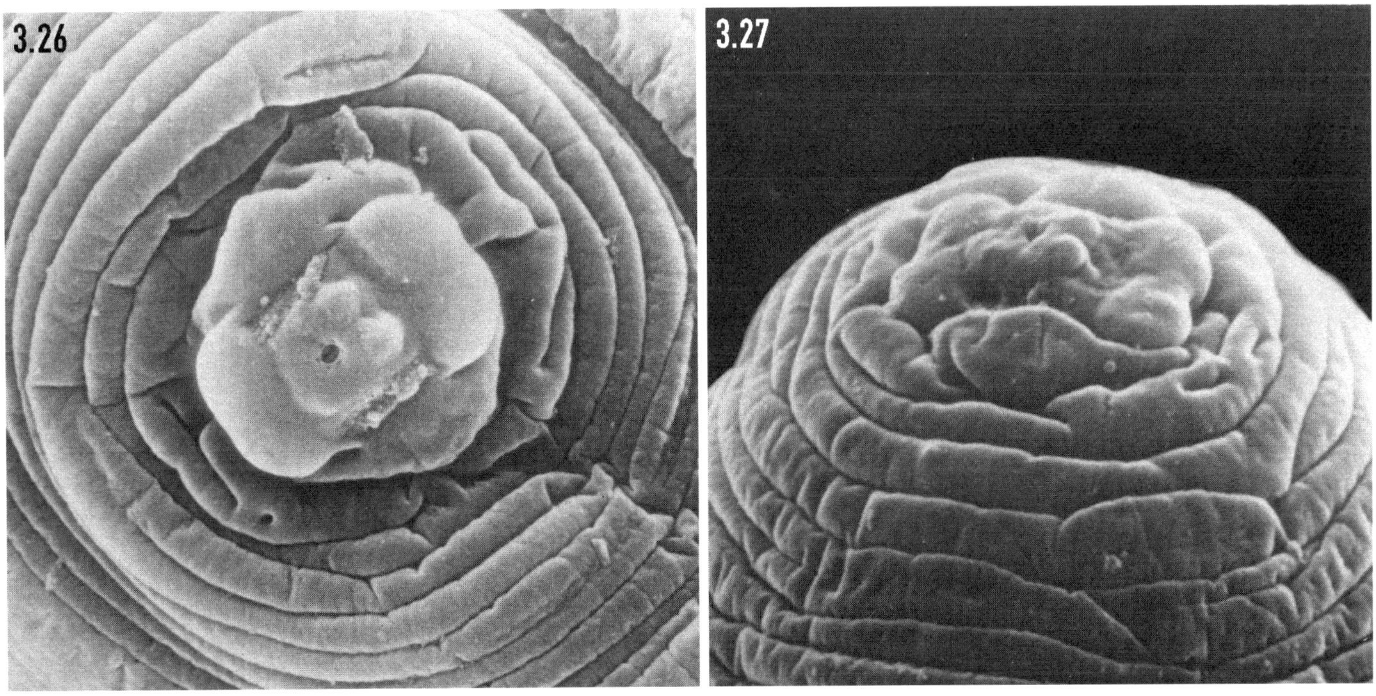

Figs. 3.26-3.27. *Meloidogyne javanica* female, face and lateral view, respectively (SEM)

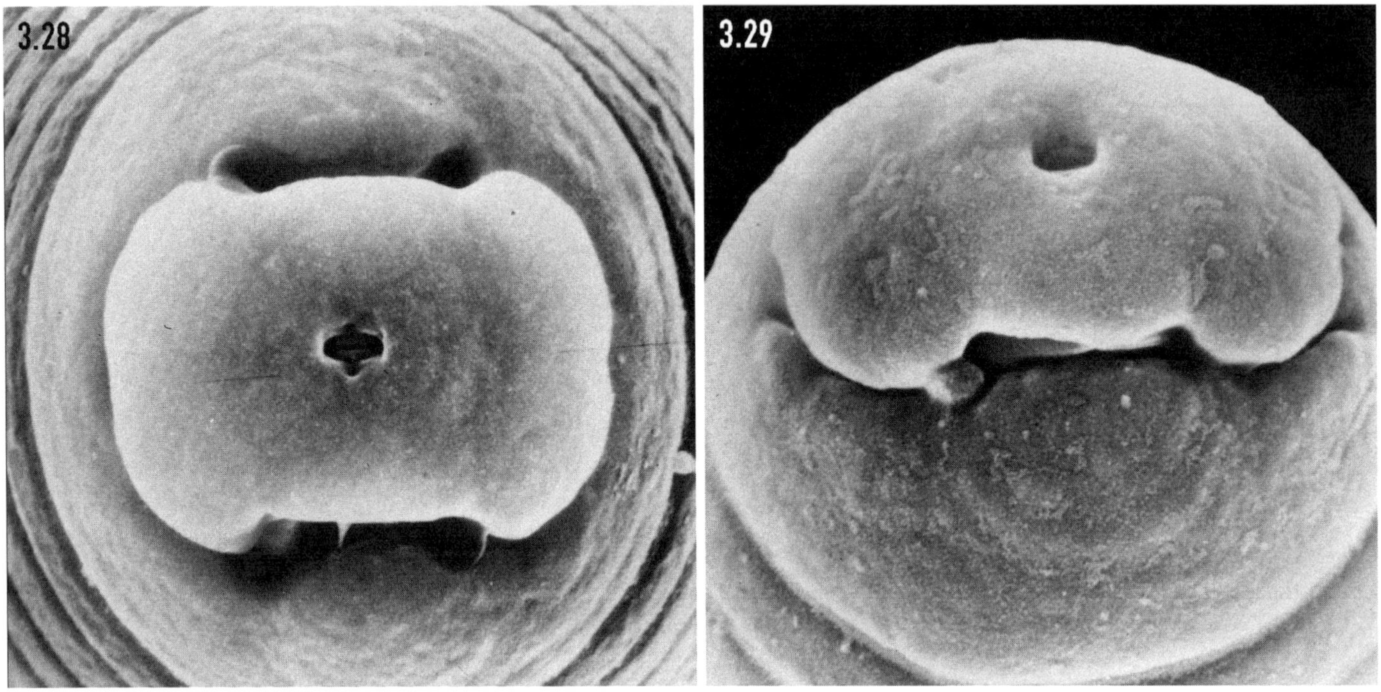

Figs. 3.28-3.29. *Meloidogyne javanica* male, face and lateral view, respectively (SEM)

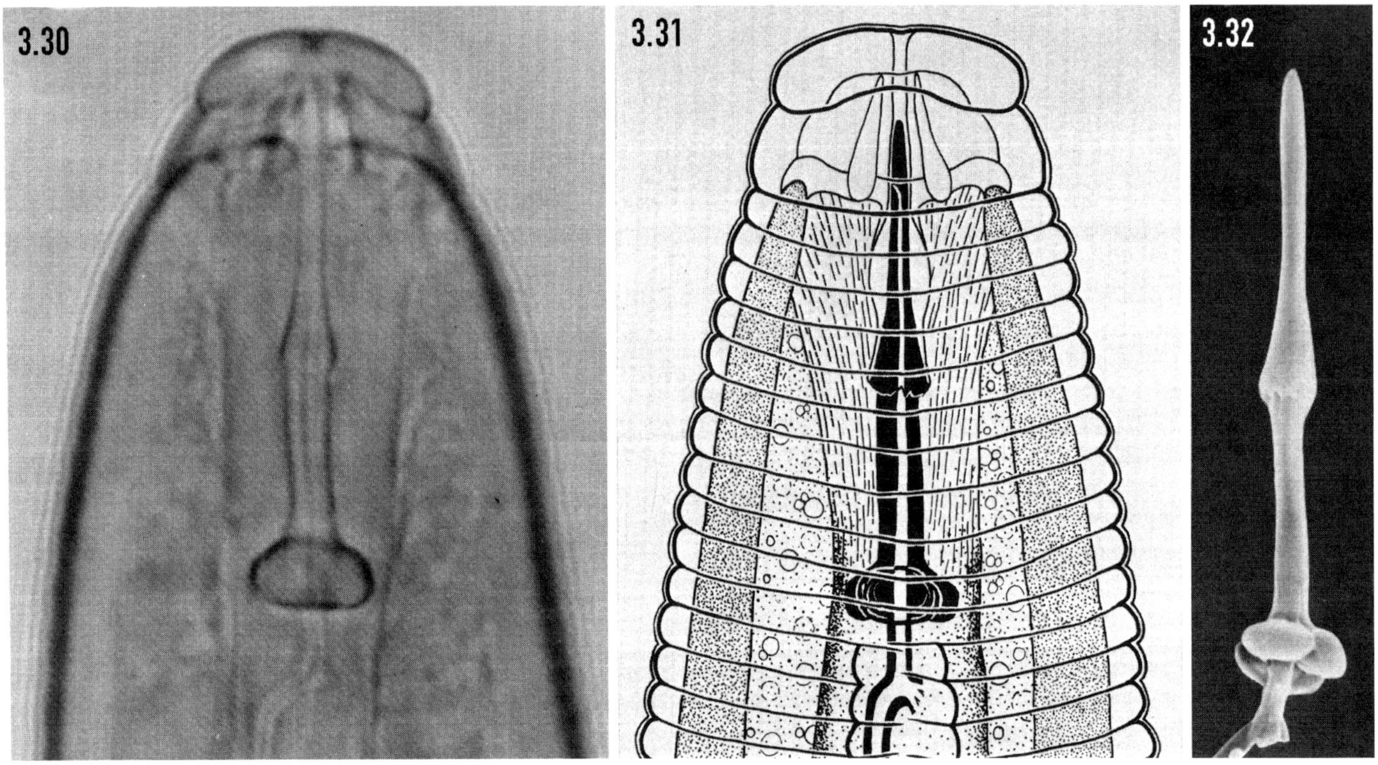

Figs. 3.30-3.31. LM photograph and line drawing of the head and stylet of a *Meloidogyne javanica* male. **3.32.** SEM photograph of an excised stylet of a male of *M. javanica*

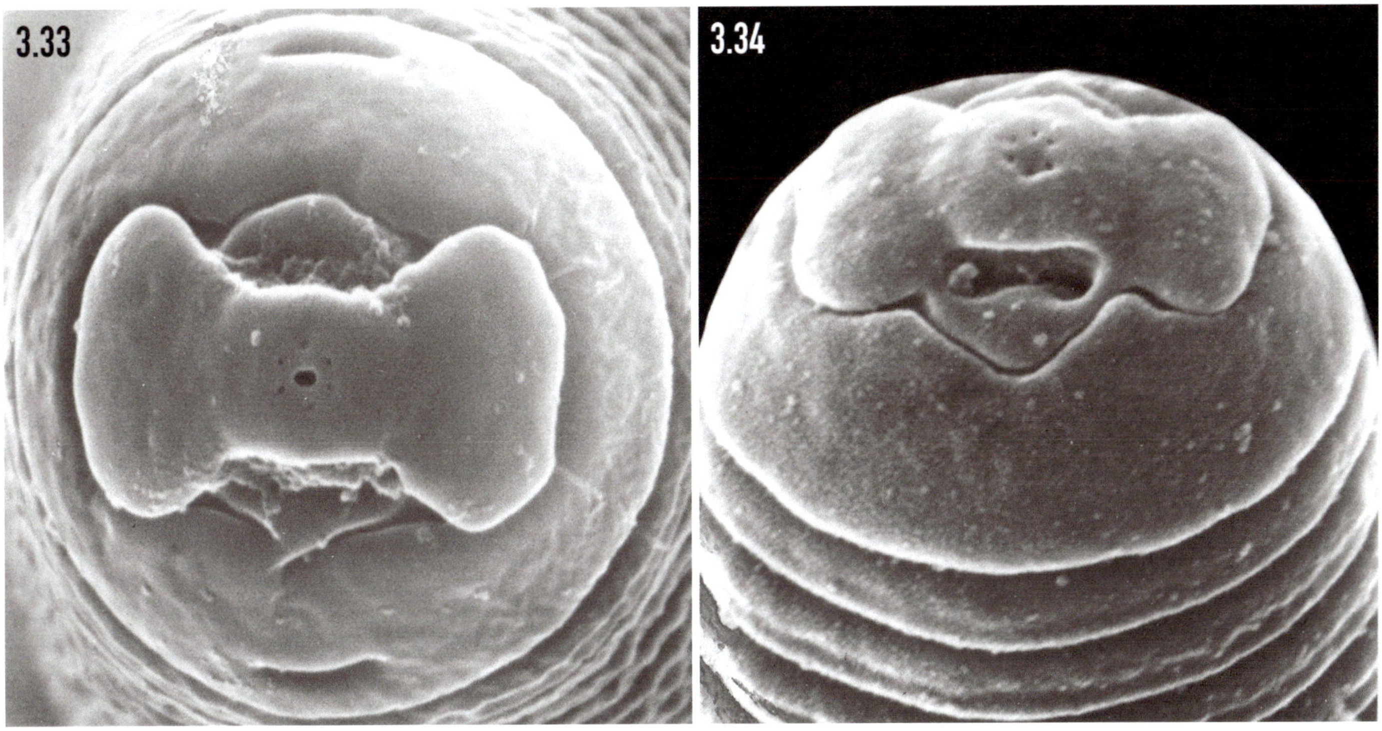

Figs. 3.33-3.34. Second-stage juvenile of *Meloidogyne javanica*, face and lateral view, respectively (SEM)

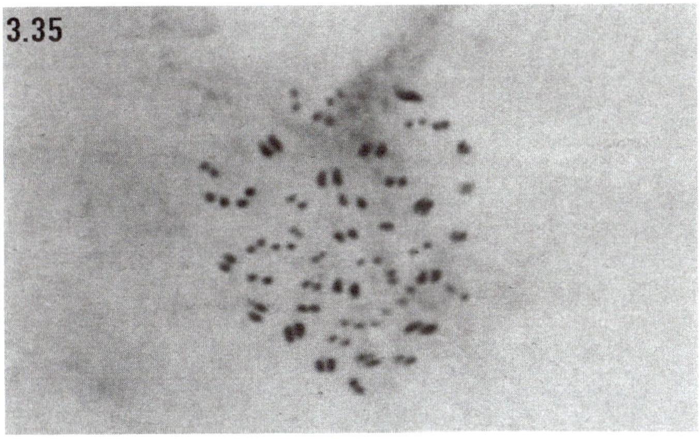

Fig. 3.35. Prometaphase chromosomes of the single maturation division of oocytes of *M. javanica*. The chromosomes are univalent (dyads), an indication that the species reproduces by mitotic parthenogenesis (After Triantaphyllou: Nematologica 7:105-113, 1963)

III. *Meloidogyne arenaria*

A. Morphology

All photographs are from IMP population 351 from Florida, host race 1 with 54 chromosomes.

1. Females

a. **Perineal patterns** - The dorsal arch in *M. arenaria* populations **(Figs. 3.36-3.39)** is flattened to rounded. The striae in the arch are slightly indented at the lateral lines and generally form a shoulder on the arch. Often the dorsal and ventral striae meet at an angle at the lateral lines. Some striae fork and are short and irregular near the lateral lines. The striae are smooth to wavy and some may bend toward the vulva. The patterns may also have striae that extend laterally to form one or two wings. Variants of some populations are similar to patterns of *M. hapla*, or *M. incognita*.

b. **Stylets** - *M. arenaria* females **(Figs. 3.40-3.42)** have unique stylets that are very characteristic for the species. In general the stylet is very robust; both the cone and shaft are broad. The shaft increases in width posteriorly and gradually merges with the stylet knobs. The knobs are wide and rounded posteriorly.

c. **Head morphology** - The labial disc and medial lips of *M. arenaria* **(Figs. 3.43-3.44)** are dumb bell shaped. Like *M. incognita* and *M. javanica*, the lateral lips are large and separated from the medial lips and head region. Usually the head region has one incomplete annulation.

M. arenaria

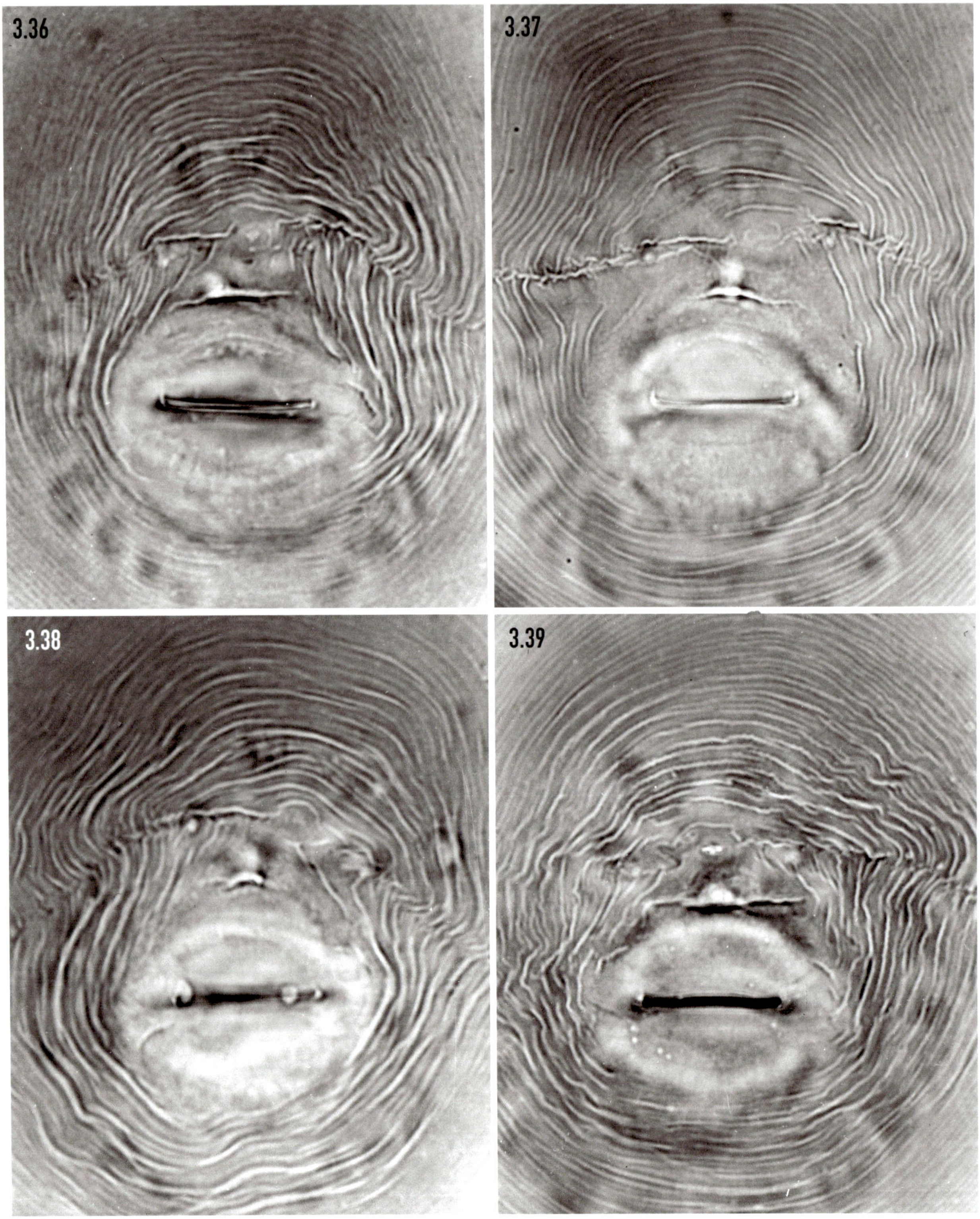

Figs. 3.36-3.39. Perineal patterns of *Meloidogyne arenaria*

2. Males

a. **Head morphology** - The head cap of *M. arenaria* males **(Figs. 3.45-3.48)** is low and slopes posteriorly. It forms a smooth and continuous structure that is almost as wide as the head region. Two or three incomplete annulations are present on the head region.

b. **Stylets** - The stylet cone of *M. arenaria* males **(Figs. 3.47-3.49)** is pointed and the lumen opening is marked on the ventral side by a slight protuberance. The posterior portion of the cone is much wider than the anterior portion of the shaft. Generally, the shaft is cylindrical, although its diameter

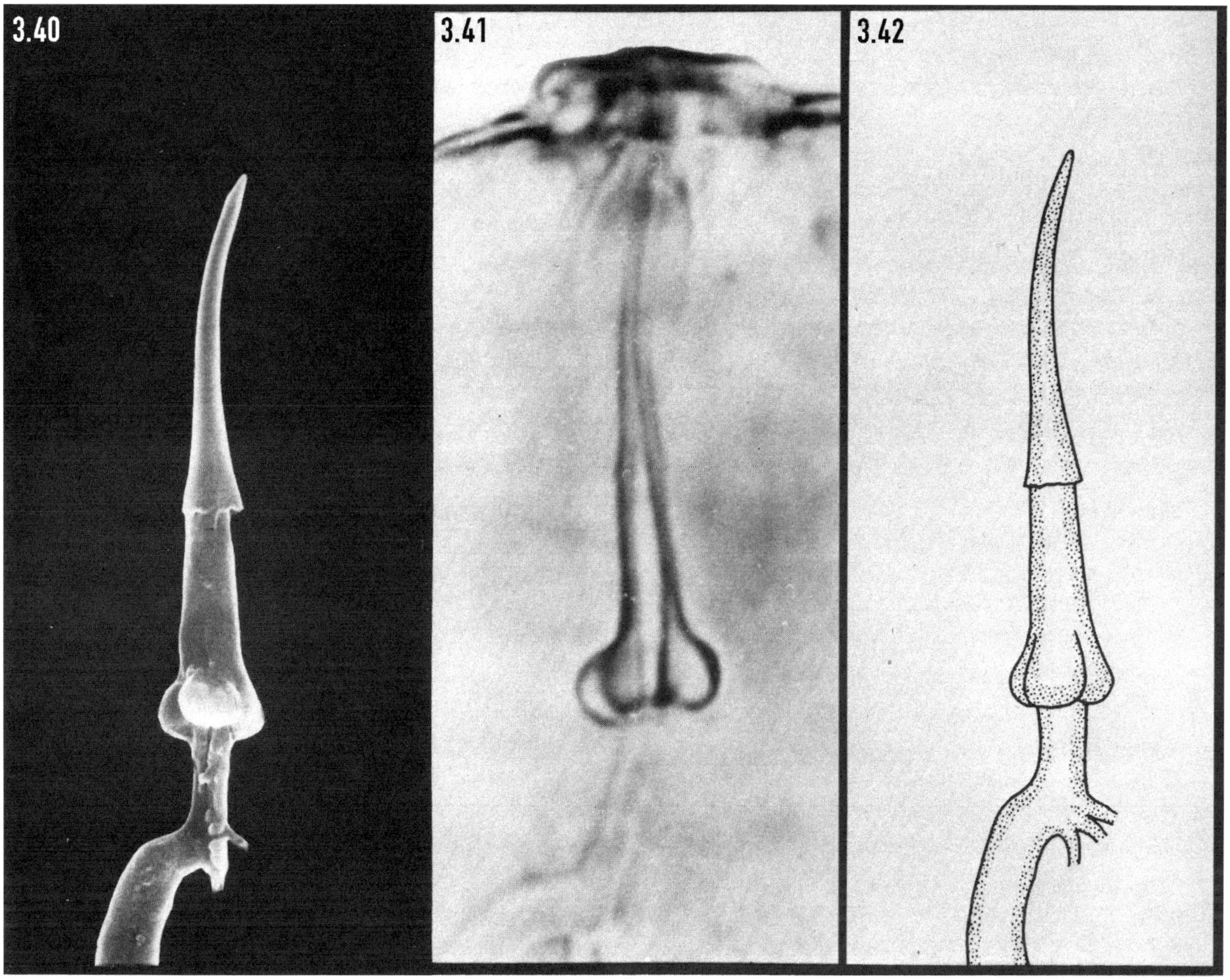

Figs. 3.40-3.42. SEM and LM photographs and line drawing of a stylet of a female of *Meloidogyne arenaria*

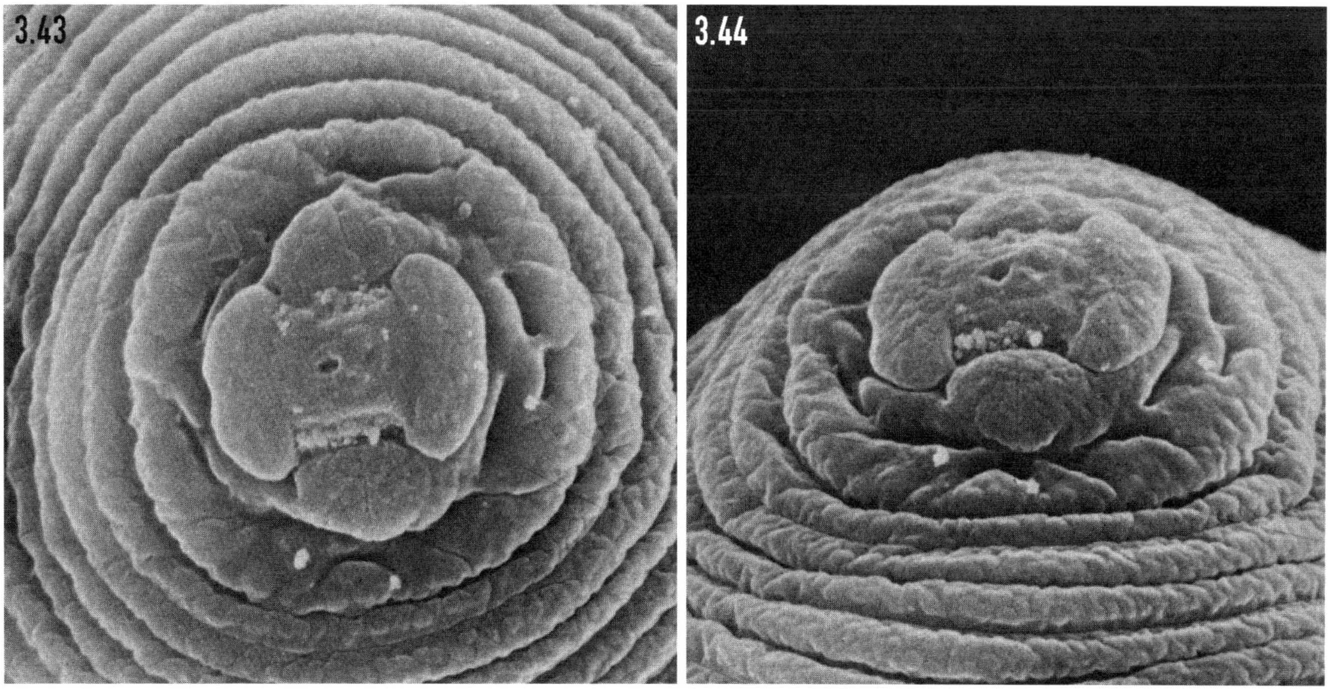

Figs. 3.43-3.44. *Meloidogyne arenaria* female, face and lateral view, respectively (SEM)

may increase slightly medially. The anteriorly indented knobs are very large and gradually merge with the shaft.

2. Males

a. **Head morphology** - The head cap of *M. arenaria* males **(Figs. 3.45-3.48)** is low and slopes posteriorly. It forms a smooth and continuous structure that is almost as wide as the head region. Two or three incomplete annulations are present on the head region.

b. **Stylets** - The stylet cone of *M. arenaria* males **(Figs. 3.47-3.49)** is pointed and the lumen opening is marked on the ventral side by a slight protuberance. The posterior portion of the cone is much wider than the anterior portion of the shaft. Generally the shaft is cylindrical, although its diameter may increase slightly medially. The anteriorly indented knobs are very large and gradually merge with the shaft.

3. Second-Stage Juveniles

Head morphology - In *M. arenaria* **(Figs. 3.50- 3.51)**, the labial disc and medial lips are dumb bell shaped and elongate. The lateral lips are long and lie below the labial disc and medial lips. In most specimens the head region is not annulated, although some specimens do have two or three head annulations.

4. Useful Measurements

(from three populations, 30 specimens each)

Second-stage juvenile total length, 398-605 (521) µm; range (mean); tail length, 44-69 (58) µm; head end to stylet base, 14-16 (15) µm; female stylet length, 13-17 (15) µm; male stylet length, 20-25 (22) µm.

B. Differential Host Test

Populations of *M. arenaria* are separated by the North Carolina Differential Host Test into two races. Race 1 populations reproduce on peanuts, but race 2 populations do not. Most populations of both races reproduce on resistant tobacco, watermelon, and tomato but not on cotton.

C. Symptomatology

Meloidogyne arenaria populations often produce many small bead-like galls that do not form short lateral roots. A photograph of typical root galling is presented in **Fig. 3.52**. Quite often, however, galls produced by *M. arenaria* populations may be similar to those of *M. incognita* and *M. javanica*.

D. Cytogenetics

All populations of *M. arenaria* reproduce by mitotic parthenogenesis. Two chromosomal races are recognized in this species. Race A is the most common and includes triploid popula-

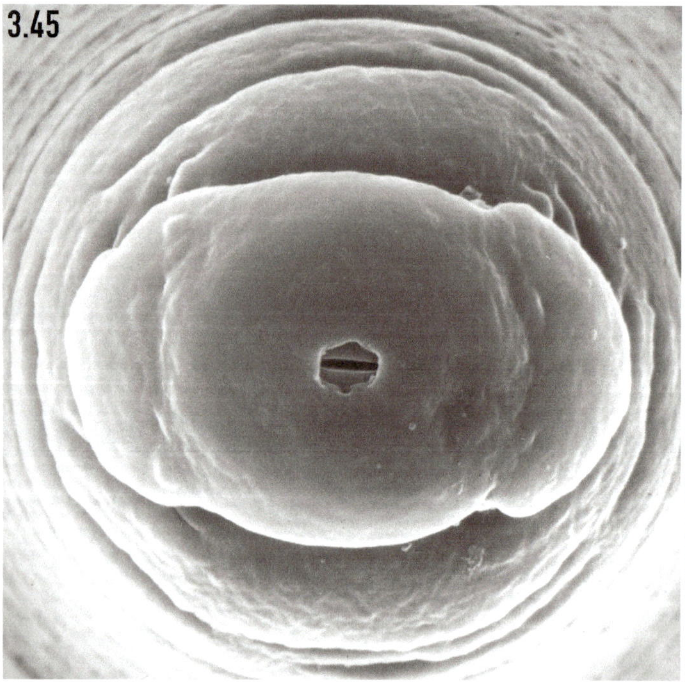

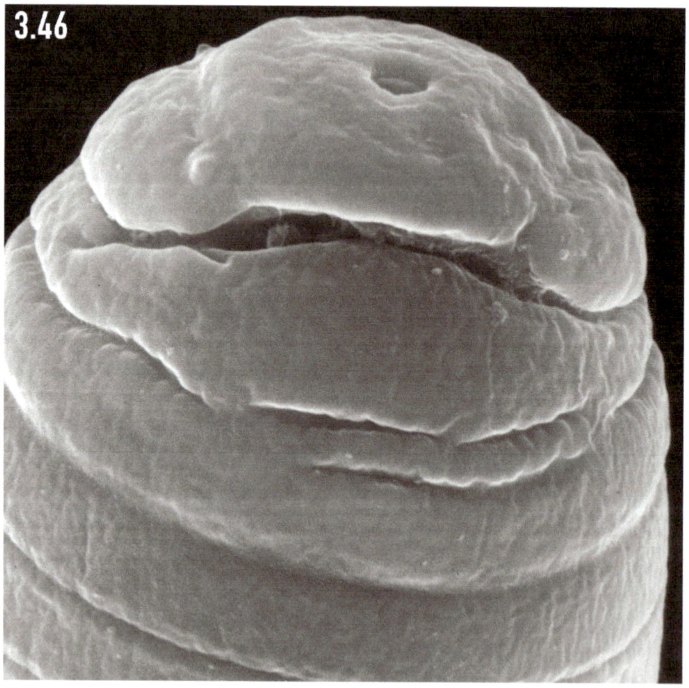

Figs. 3.45-3.46. Male of *Meloidogyne arenaria*, face and lateral view, respectively (SEM)

tions with 2n=50 to 56 chromosomes **(Fig. 3.53)**. Race B is the diploid race with 2n=34 to 37 **(Fig. 3.54)**. The chromosomes of *M. arenaria* are similar in morphology and behavior to those of M. javanica. The two species differ only in chromosome numbers. Therefore, determination of the

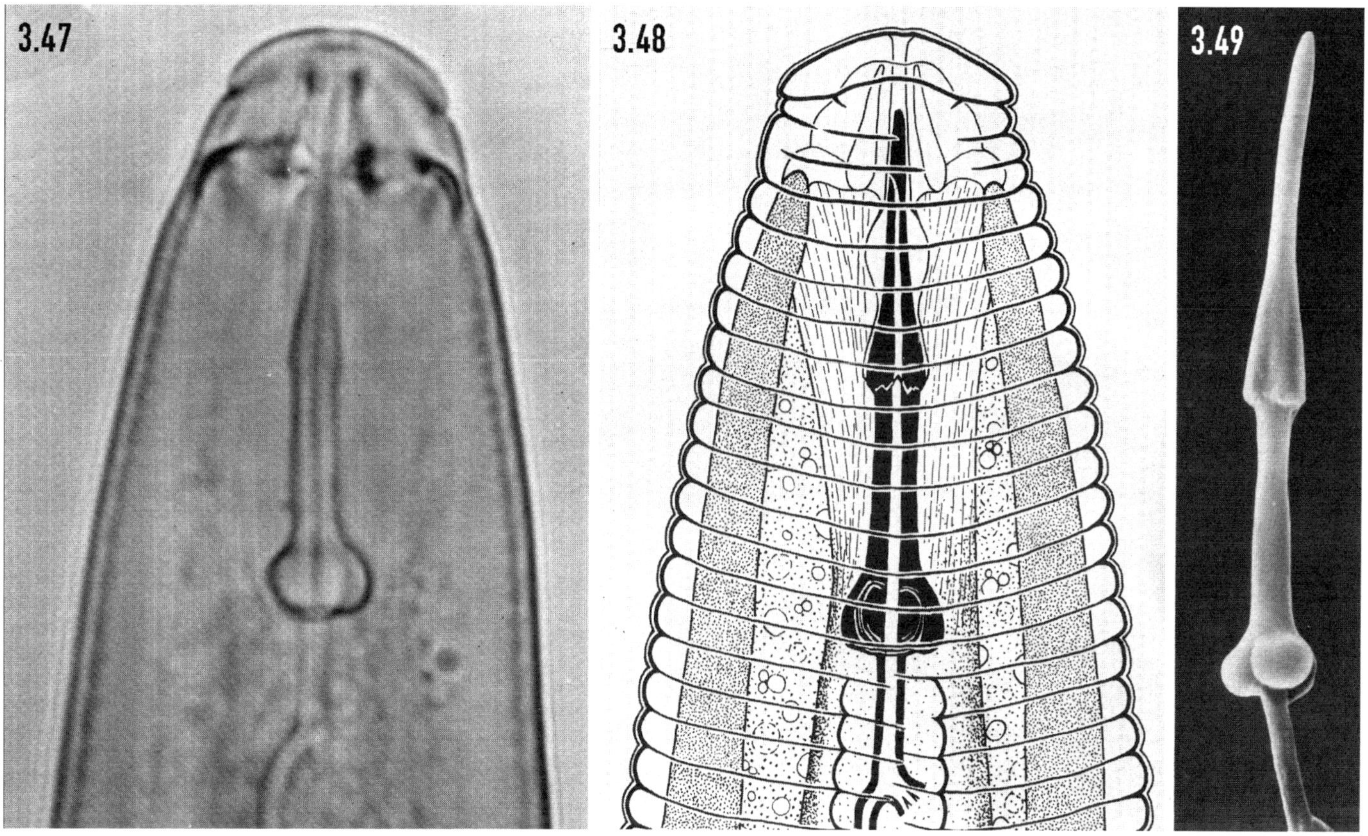

Figs. 3.47-3.48. LM photograph and line drawing of the head and stylet of a male of *Meloidogyne arenaria*. **3.49.** SEM photograph of an excised stylet of a male of *M. arenaria*

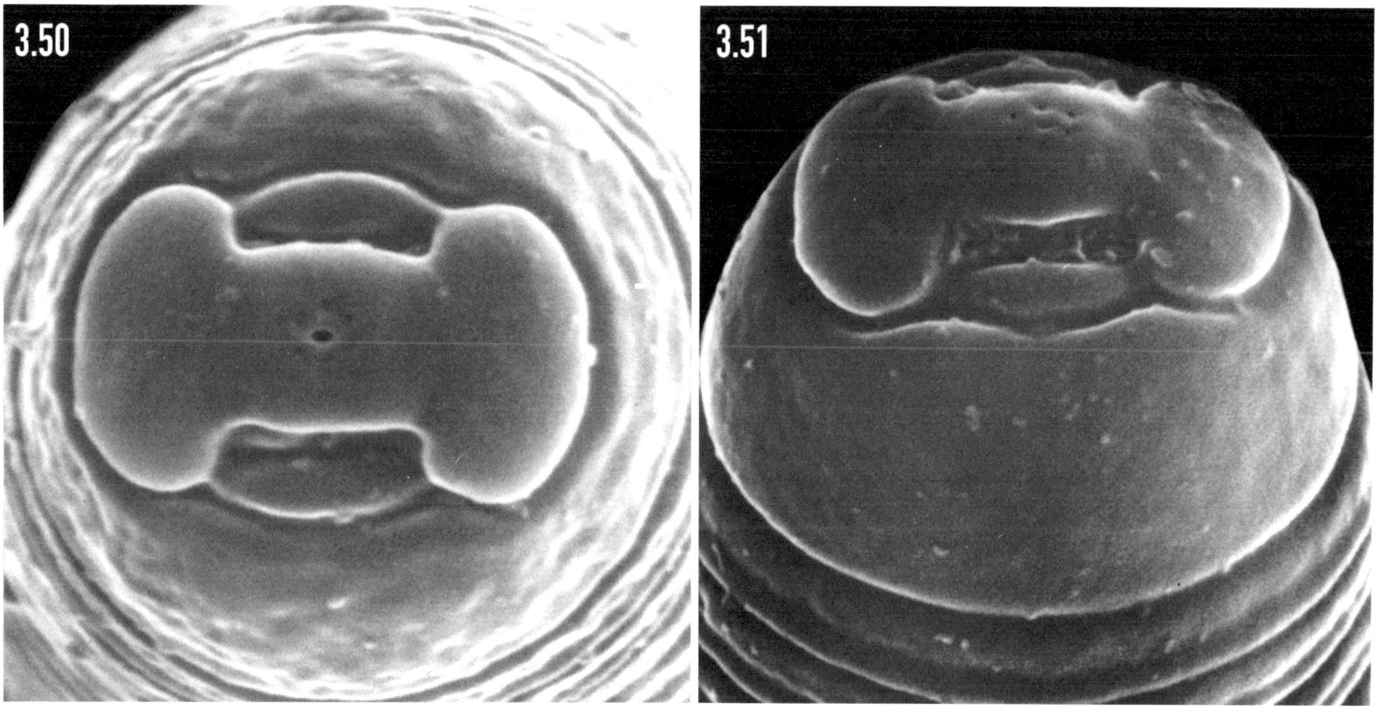

Figs. 3.50-3.51. Second-stage juvenile of *Meloidogyne arenaria*, face and lateral view, respectively (SEM)

approximate chromosome number is essential for differentiating between these two species.

E. Biochemistry

There are two forms of *M. arenaria* with reference to esterase patterns **(Fig. 1.10)**. The most common form has two major bands of esterase activity at Rf=0.54 and 0.57. The other form has, in addition, a band at Rf=0.50.

F. Ecology

Meloidogyne arenaria is considerably less important than *M. incognita* or *M. javanica*, constituting about 8% of the populations encountered in samples collected through the IMP. Distribution in the northern and southern hemispheres approximates that of *M. incognita*.

The optimum warm month temperature is approximately 24°C. Two host races are recognized. Race 1 reproduces on peanut; race 2 does not. Race 2 frequently will not reproduce on pepper, and neither race reproduces on cotton or strawberry.

3.52

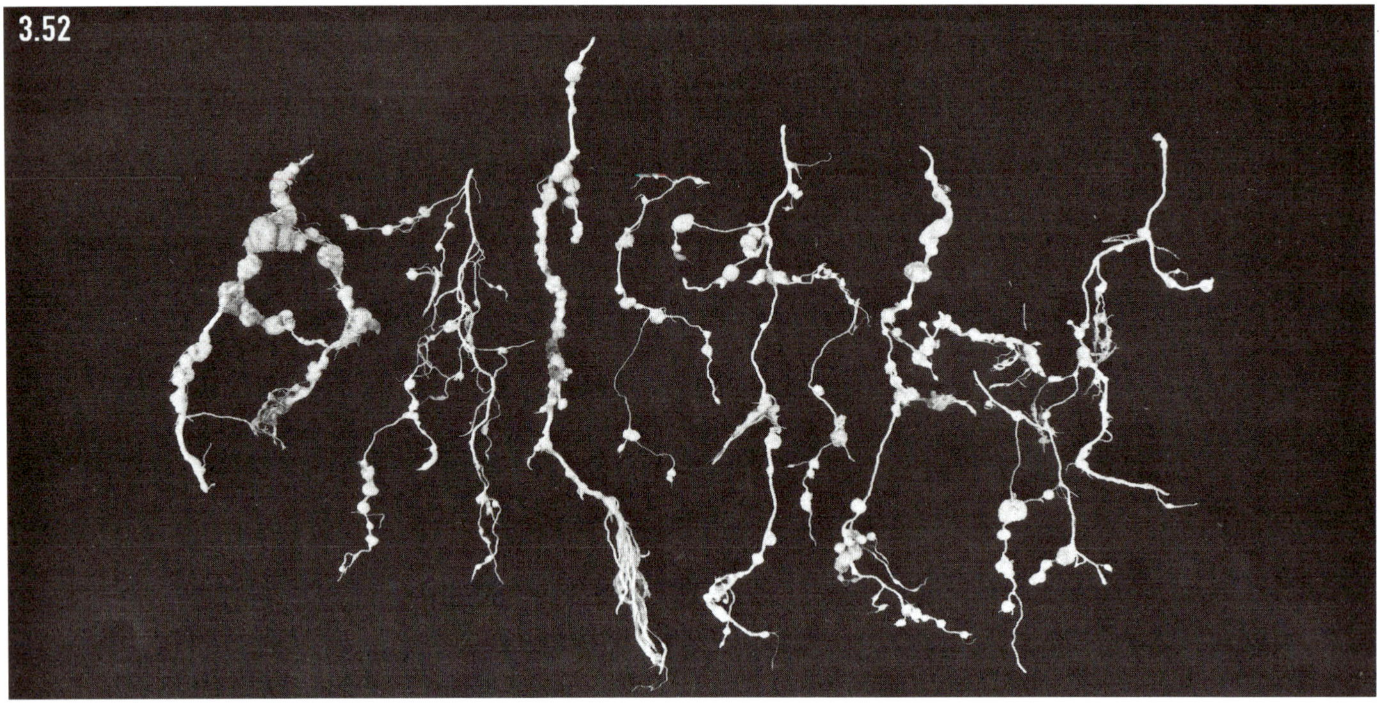

Fig. 3.52. **Tomato roots with galls caused by *Meloidogyne arenaria***

3.53

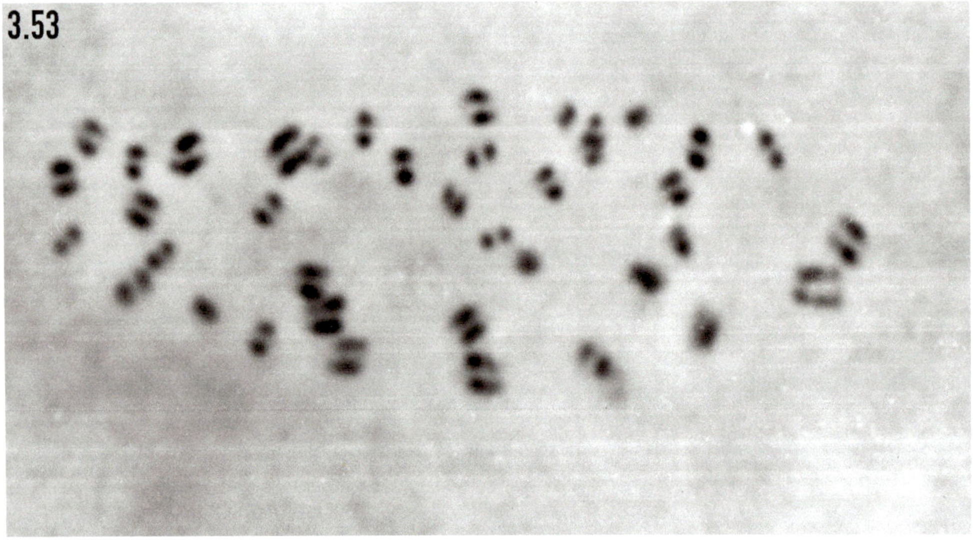

3.54

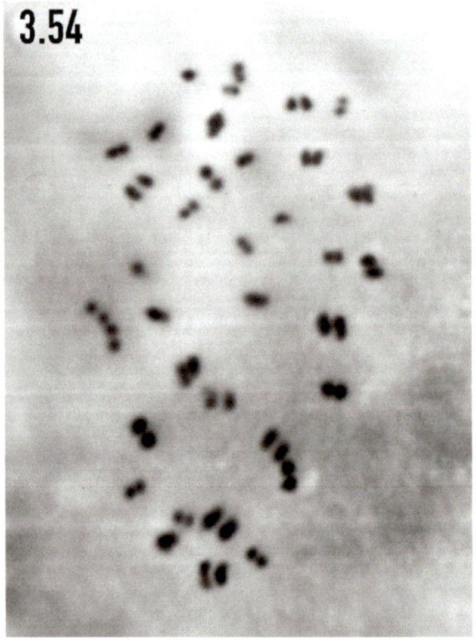

Figs. 3.53-3.54. Prometaphase chromosomes of the single maturation division of oocytes of M. arenaria. 3.53) Triploid race A with 53 chromosomes. 3.54. Diploid race B with 36 chromosomes. Both races have univalent chromosomes (dyads) and reproduce by mitotic parthenogenesis (After Triantaphyllou: J. Morphol. 113:487-499, 1963)

IV. *Meloidogyne hapla*

A. Morphology

Photographs are from IMP population 42 from Canada with 15 chromosomes; populations 6, 86, and 48 from North Carolina with 16, 17, and 45 chromosomes, respectively; population 66 from Maryland with 45 chromosomes and population 230 from Chile with 48 chromosomes

1. Females

a. **Perineal patterns** - Patterns of *Meloidogyne* populations **(Figs. 3.55-3.58)** are nearly round hexagons to slightly flattened ovals. The dorsal arch is generally flattened. Lateral lines are indistinct, although they may be indicated by slight irregularities in the striae or by dorsal and ventral striae that meet at an angle. Some striae may extend laterally and form one or two wings. Striae are smooth to wavy. The tail terminal area is usually marked by punctations which is a good character for this species. Punctations may be absent in some preserved specimens due to fixation procedures.

b. **Stylets** - The stylets of *M. hapla* females **(Figs.3.59-3.61)** are small compared to the other three common species. The cone is only slightly curved dorsally and the shaft is broadest posteriorly. *M. hapla* stylet knobs are rounded and distinctly set off from the shaft.

c. **Head morphology** (SEM) - The labial disc and medial lips of M. hapla females **(Figs. 3.62-3.67)** are asymmetric. The small triangular lateral lips fuse with the ventral lip but are set off from the dorsal lip. The head region is large and not annulated. The head morphology of various cytological populations of race A and B is very similar, even though certain details are different. The population of race A with 15 chromosomes **(Figs. 3.62-3.63)** is different from the typical *M. hapla* populations in that the medial lips are pointed.

2. Males

a. **Head morphology** - Males from race A and B of *M. hapla* have similar head morphology **(Figs. 3.68-3.77, 3.79-3.80)**. The head cap is high and much narrower than the head region. The non-annulated head region is generally set off from the body annules because its diameter is larger than that of the first body annule. Also, the body annules decrease in width and height as they near the head region. Differences between the races of *M. hapla* are slight. Race A males **(Figs. 3.68-3.73)** have an indication of lateral lips, whereas race B males **(Figs. 3.74- 3.75)** lack lateral lips. One population of race A with 15 chromosomes **(Figs. 3.68-3.69, 3.82-3.83)** has unique head morphology. The head cap is lower and wider and the medial lips are pointed instead of squared-off or rounded. Also, the head region is not set off from the body annules which are fainter than in the other populations. It is difficult to determine by LM where the head region ends and the body annules begin. The stylet morphology of this population, however **(Figs. 3.82-3.83),** is typical for race A of *M. hapla*.

b. **Stylets** - The stylets of *M. hapla* **(Figs. 3.76-3.83)** are much thinner and shorter than those of the other three common species. The cone gradually increases in width posteriorly and the base of the cone is not much wider than the anterior portion of the shaft. The shaft of race A populations broadens as it nears the stylet knobs. The knobs are round and set off from the shaft. The stylets of race B populations **(Figs. 3.79-3.81)** are different from race A. They are longer, the shaft remains cylindrical and often indents at its junction with the stylet knobs, and the knobs are larger. The two races can, therefore, be separated on the basis of stylet morphology.

3.Second-Stage Juveniles

Head morphology (SEM) - Some differences occur in the head morphology of second-stage juveniles between race A and B of *M. hapla* **(Figs. 3.84-3.91)**. Race A populations with different chromosome numbers **(Figs. 3.84-3.89)** also differ in some respects. Race B populations **(Figs. 3.90-3.91)** are, on the other hand, uniform morphologically. In all populations of race A the labial disc is fused with the medial lips and in the same contour. Differences between the chromosomal populations occur in medial lip shape. The population with 15 chromosomes has pointed medial lips, the medial lips of the population with 16 chromosomes are rectangular, and the population with 17 chromosomes has rounded medial lips. In all populations of race B, the labial disc is rounded and raised above the rounded medial lips. The head region is smooth in all populations of *M. hapla*.

M. hapla

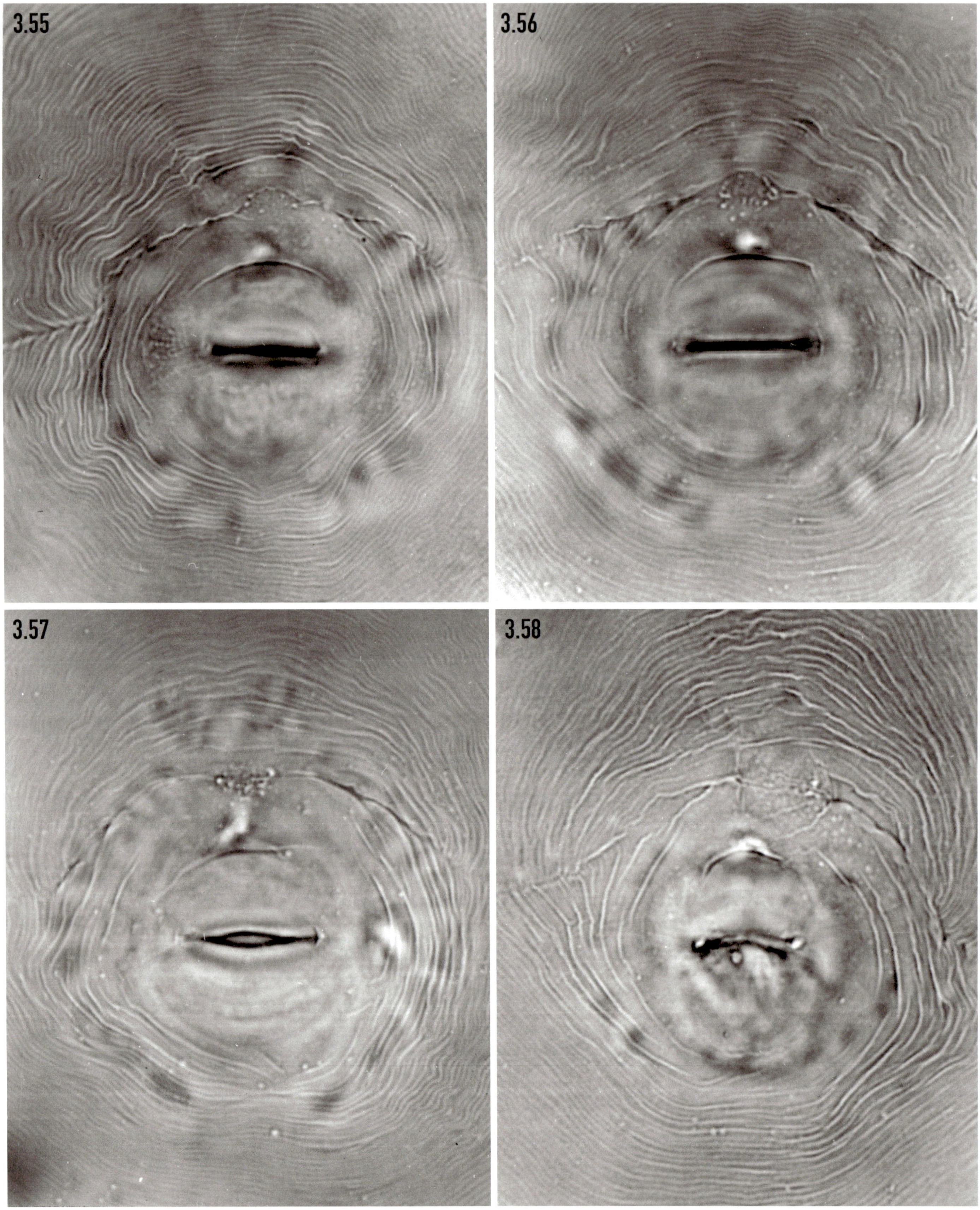

Figs. 3.55-3.58. Perineal patterns of *Meloidogyne hapla*

4. Useful Measurements

(from three populations of each race, 30 specimens each)

Second-stage juvenile total length, race A, 357- 467 (413) µm, range (mean); total length, race B, 410-517 (475) µm; tail length, race A, 46-58 (53) µm; tail length, race B, 54-69 (62) µm; head end to stylet base, race A, 14-16 (15) µm; head end to stylet base, race B, 15-17 (16) µm; female stylet length, race A, 13-17 (15) µm; female stylet length, race B, 15-17 (16) µm; male stylet length, race A, 17-23 (20) µm; male stylet length, race B, 19-23 (21) µm.

B. Differential Host Test

Populations of *M. hapla* reproduce on resistant tobacco, pepper, peanuts, and tomato. Unfavorable hosts of *M. hapla* include cotton and watermelon, according to the North Carolina Differential Host Test.

C. Symptomatology

In susceptible plants, populations of *M. hapla* often produce root symptoms that are diagnostic for the species. The galls tend to be small and have many branch rootlets which make the root system thick and matted **(Fig. 3.92)**.

D. Cytogenetics

M. hapla is made up of populations belonging to two distinct cytogenetic races (A and B). Race A is the most common and includes populations that reproduce by facultative, meiotic parthenogenesis. Most of them have haploid chromosome numbers of n=17 or 16 **(Figs. 3.93-3.94)** and some have n=l5 or 14. Race B populations reproduce exclusively by mitotic parthenogenesis. Some of them are diploid with 2n=30 to 31 but most are triploid with 2n=43 to 48 chromosomes.

Populations of race A are readily identified cyto

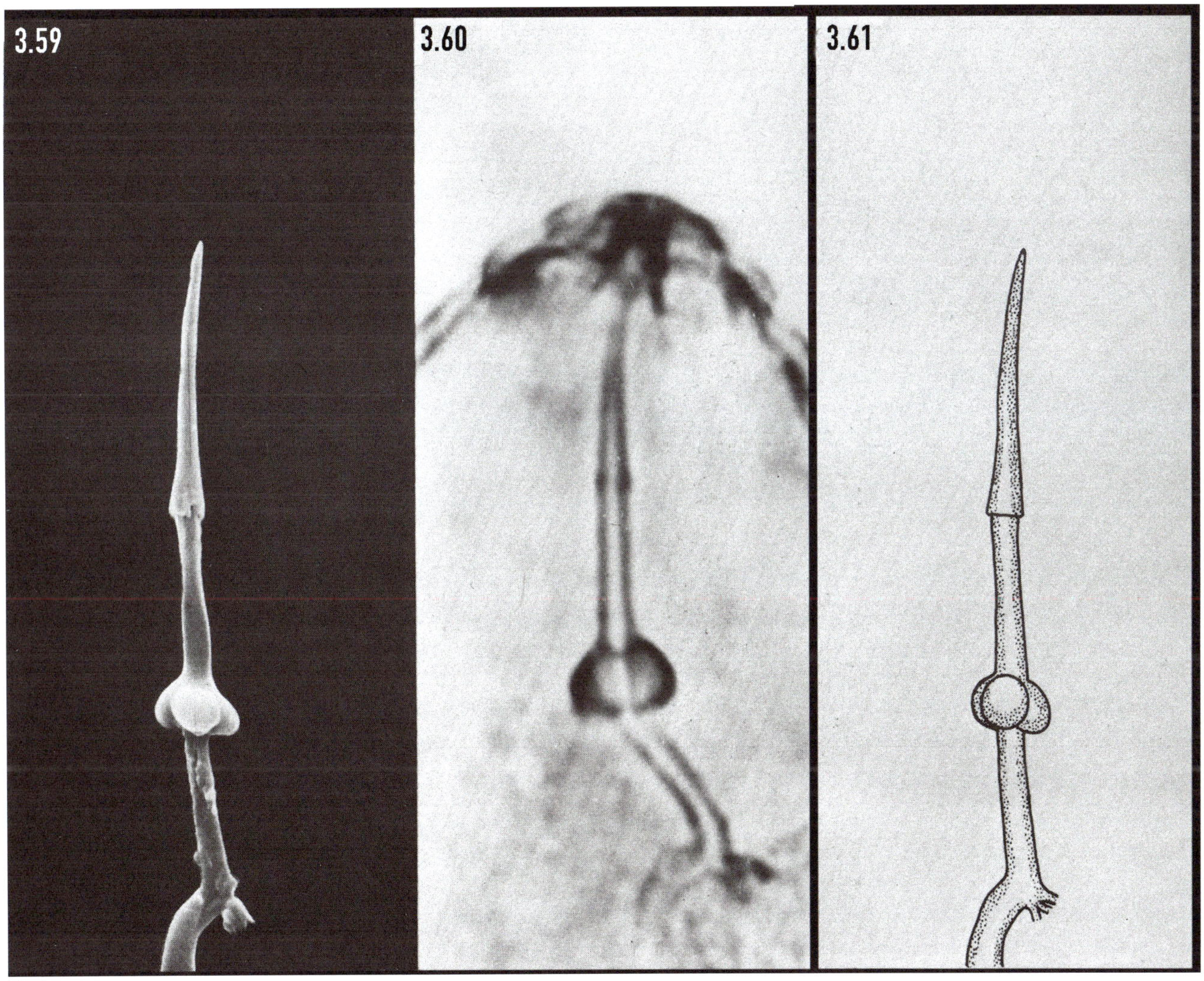

Figs. 3.59-3.61. SEM and LM photographs and line drawing of a stylet of a *Meloidogyne hapla* female

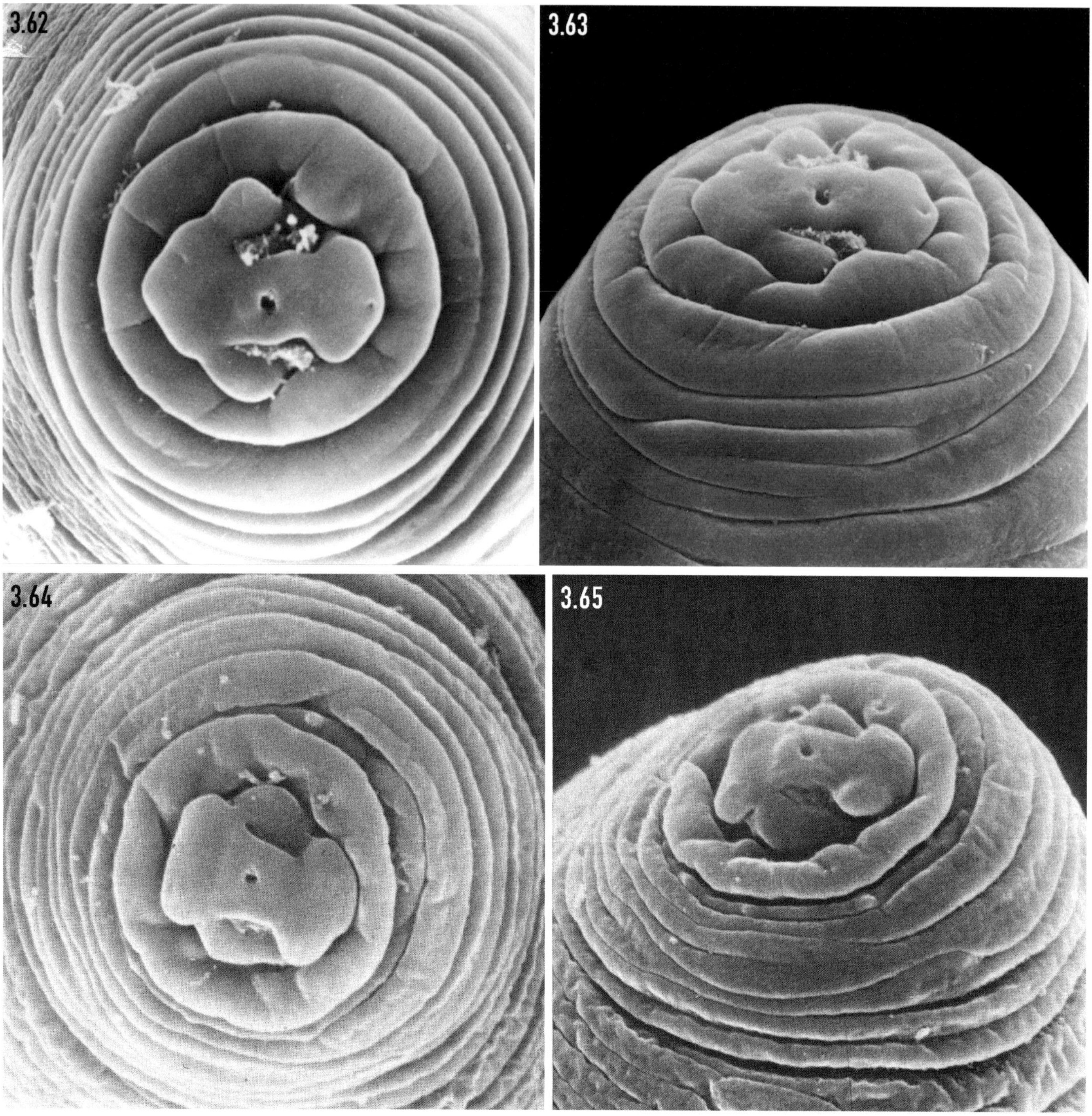

Figs. 3.62-3.65. Females of *Meloidogyne hapla* race A (SEM). **3.62- 3.63.** A female of a *M. hapla* population with 15 chromosomes, face and lateral view, respectively. **3.64-3.65.** A female of *M. hapla*, typical of populations with 16 and 17 chromosomes, face and lateral view, respectively

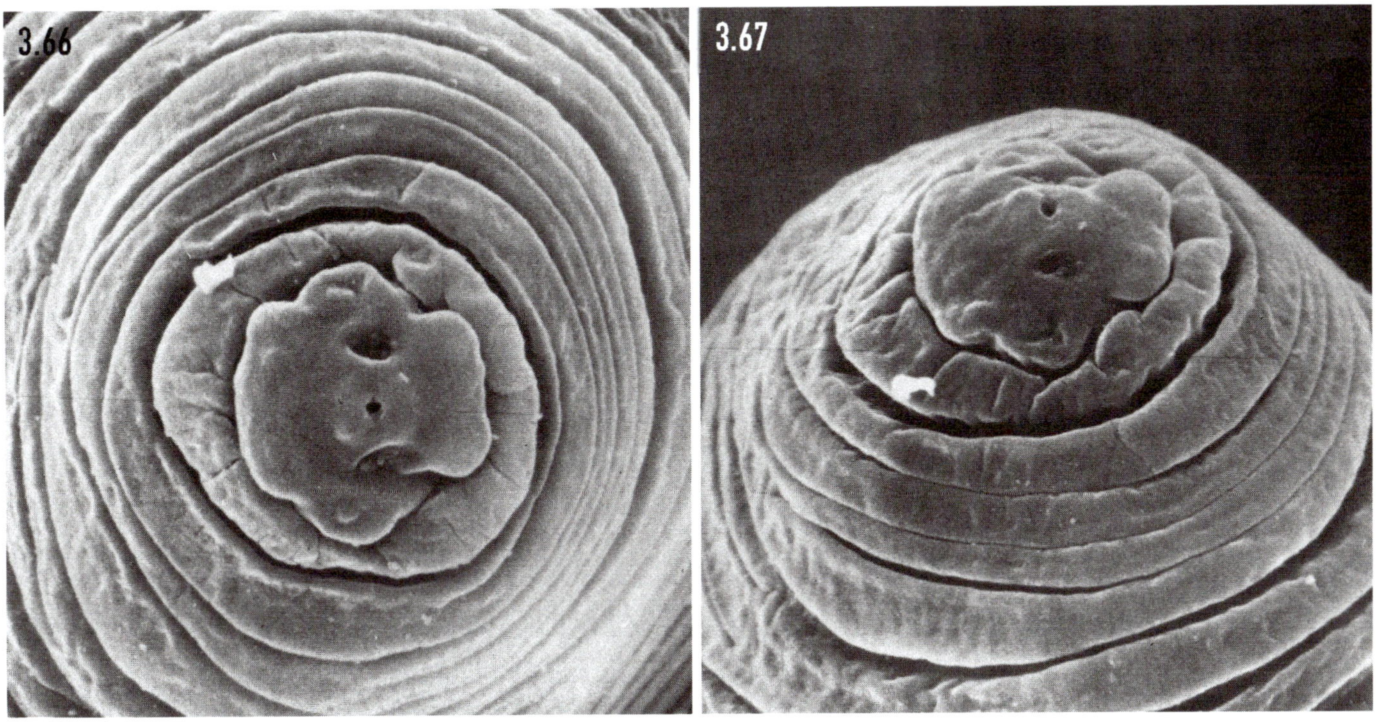

Figs. 3.66-3.67. Female of *Meloidogyne hapla* race B, face and lateral view, respectively (SEM)

genetically by the presence of 14 to 17 bivalent chromosomes (tetrads) at metaphase of the first maturation division of oocytes. None of the other three major species forms bivalents. Distinguishing race B of *M. hapla*, however, from other species is not possible without the help of other taxonomic characters. Race B populations have univalent chromosomes (dyads) similar in morphology and behavior to those of *M. arenaria* and *M. javanica*. Also, there is an overlap in chromsome numbers between *M. hapla*, race B and *M. javanica.*

E. Biochemistry

A single major band of esterase activity at Rf =0.50 is characteristic of *M. hapla* **(Fig. 1.10)**. No deviations from this pattern have been detected among 25 populations of diverse origin studied thus far. Some variation exists in minor bands which appear to have no taxonomic value.

F. Ecology

M. hapla was found in about 8% of the populations collected through the IMP. It is known to occur and cause serious losses to crops in the cooler regions of the world. In the United States, it occurs from approximately 34° N latitude to 47° N. When it occurs in the subtropical or tropical areas (Ecuador, Costa Rica, Taiwan), it is always at high elevations (1,000 or more meters). It may be found at low altitudes in the southern hemisphere north of about latitude 45° S.

Occasionally, there are reports of *M. hapla* in the southern, United States (Georgia, Florida, Louisiana) but these are believed to be recent introductions on strawberry or other plants brought in from more northern climates. *M. hapla* is more host specific than the other three species. It is a serious pest on peanuts in Virginia and North Carolina (approximately the southern limits of its occurrence in the eastern United States), as well as, strawberry, potatoes, carrots, roses, lettuce, celery and other cool climate crops.

This species does not attack watermelon, cotton, okra, or any of the grasses and grains (corn, wheat, barley, rye). Thus, a root-knot nematode attacking watermelon, cotton, okra, corn and other crops in these temperature regions would not likely be *M. hapla*.

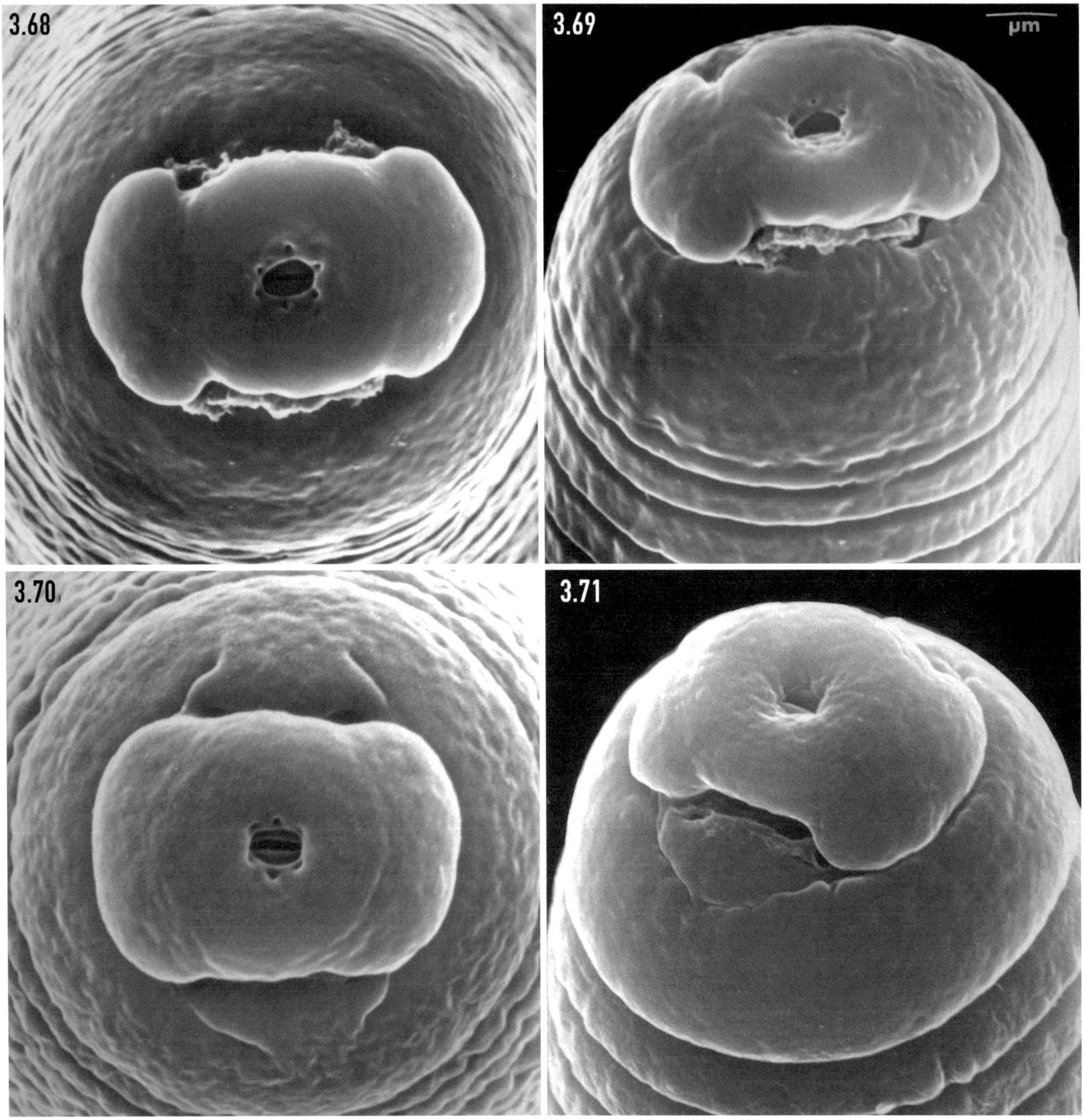

Figs. 3.68-3.71. Males of *Meloidogyne hapla* race A (SEM). **3.68-3.69.** A male of *M. hapla* population with 15 chromosomes, face and lateral view, respectively. **3.70, 3.71.** *M. hapla* male from a population with 16 chromosomes, face and lateral view, respectively

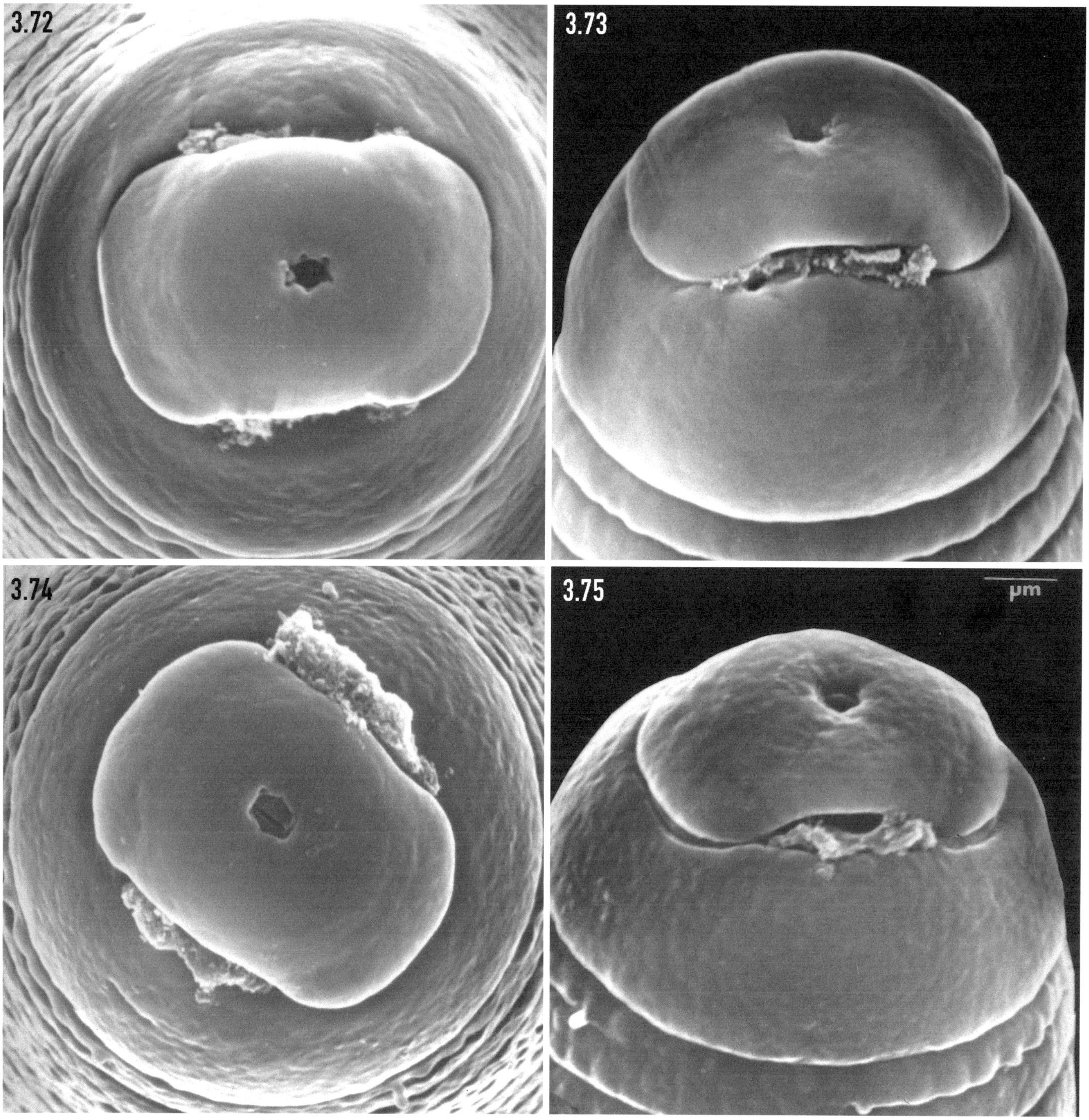

Figs. 3.72-3.75. Males of *Meloidogyne hapla* (SEM). **3.72-3.73.** A male of *M. hapla* race A population with 17 chromosomes, face and lateral view, respectively. **3.74-3.75.** A male of *M. hapla* race B population with 45 chromosomes, face and lateral view, respectively

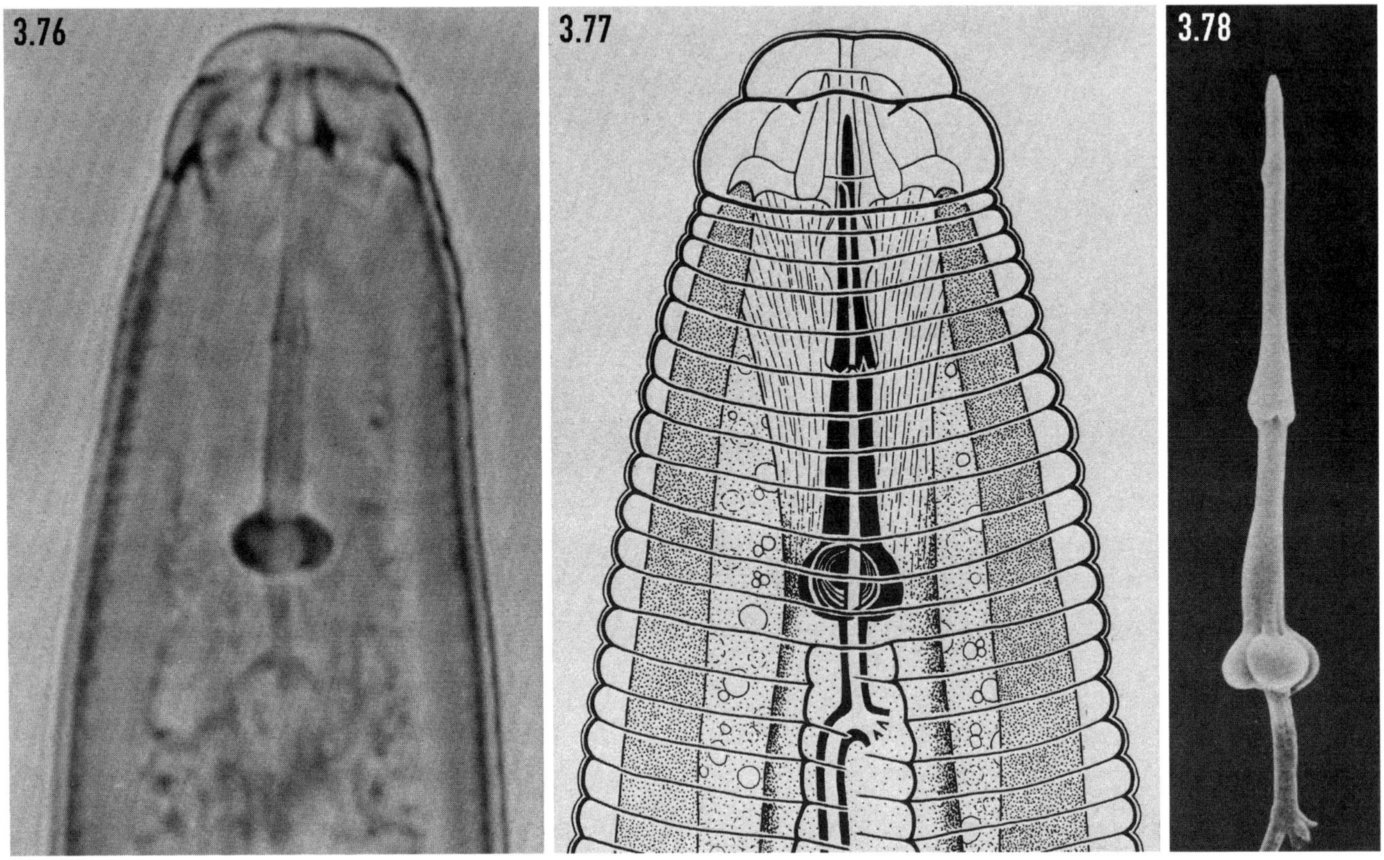

Figs. 3.76-3.78. LM photograph and line drawing of the head and stylet of a *Meloidogyne hapla* race A male. **3.78.** Excised stylet of a male of *M. hapla* race A (SEM)

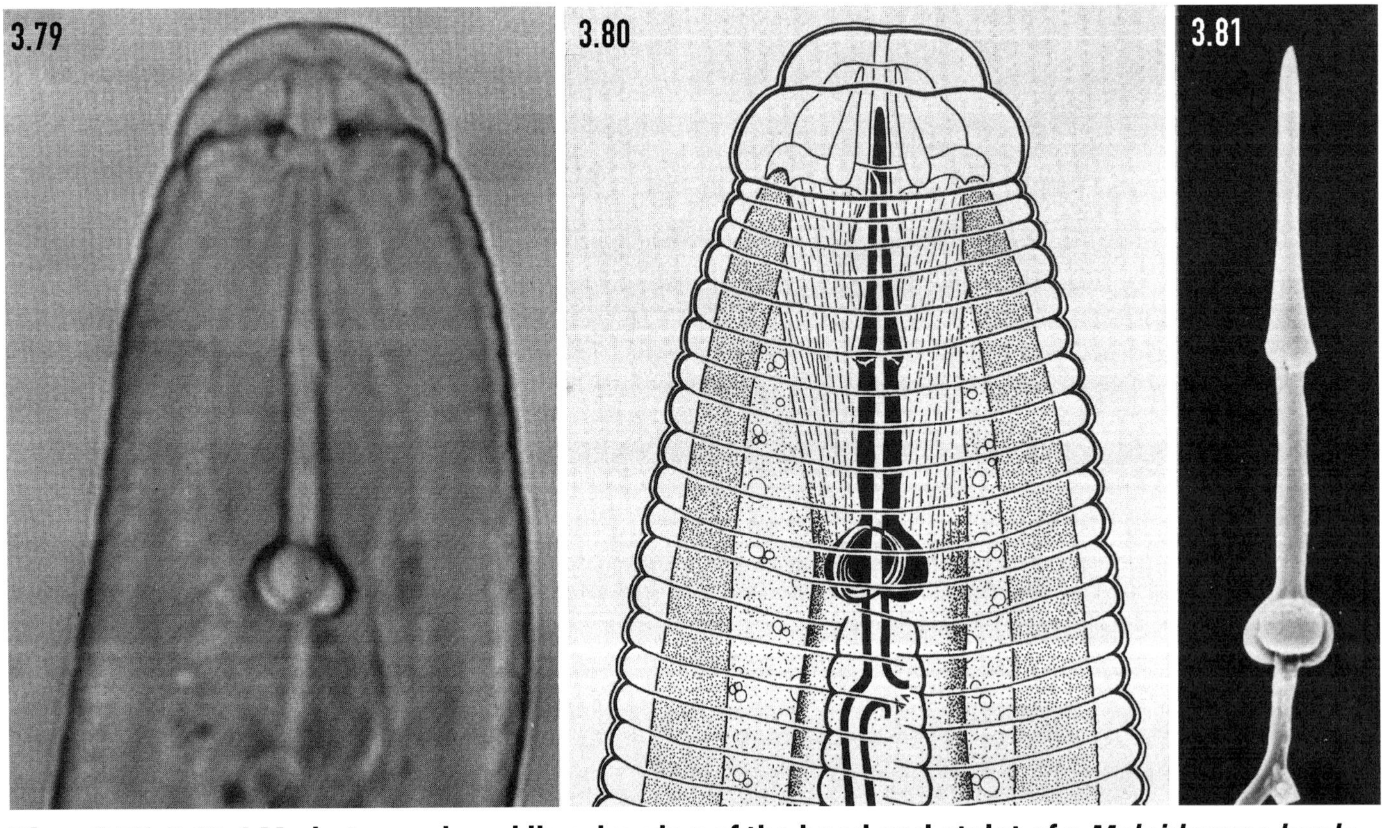

Figs. 3.79-3.81. LM photograph and line drawing of the head and stylet of a *Meloidogyne hapla* race B male. **3.81.** Excised stylet of a male of *M. hapla* race B (SEM)

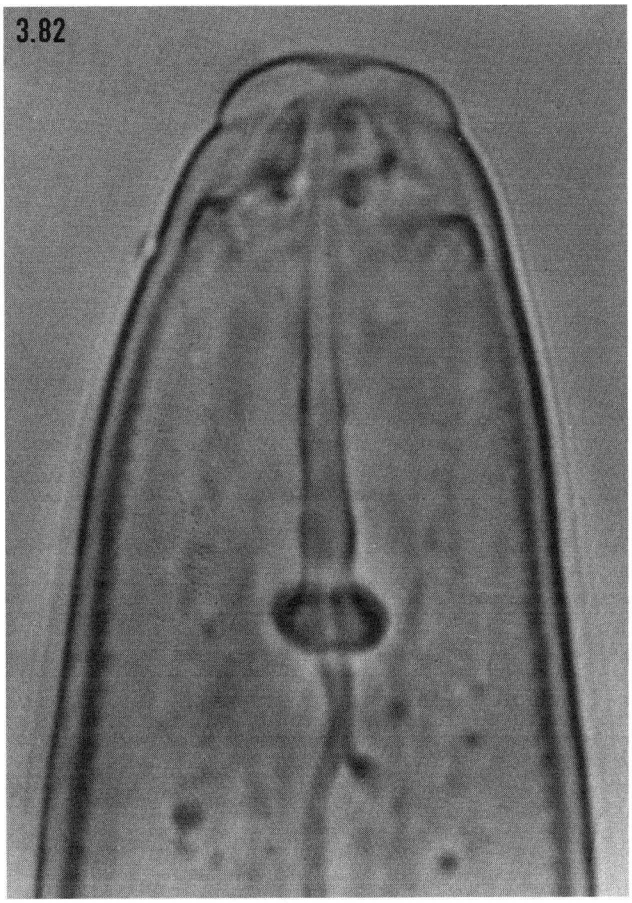

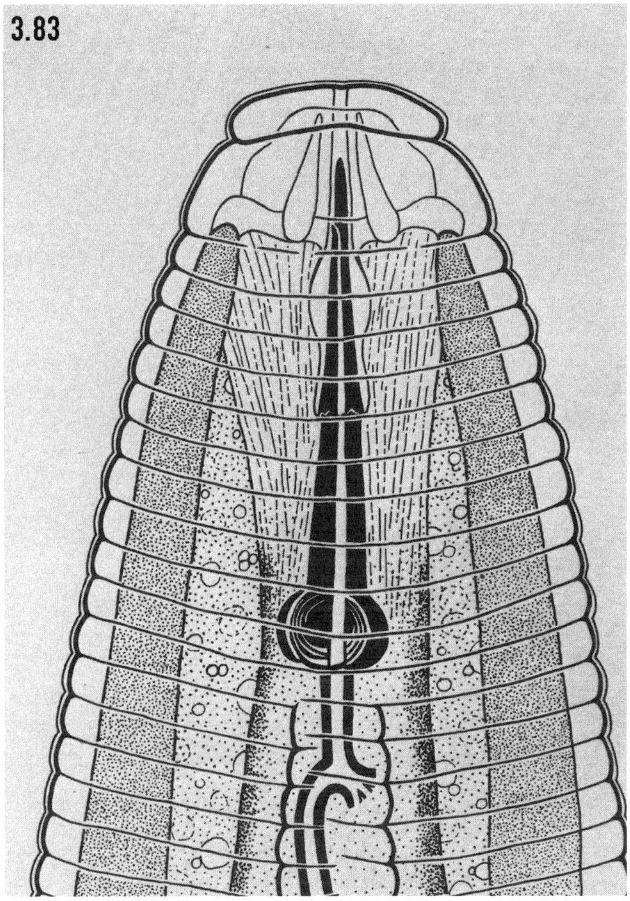

Figs. 3.82-3.83. LM photograph and line drawing of the head and stylet of a male of *Meloidogyne hapla* race A, population with 15 chromosomes

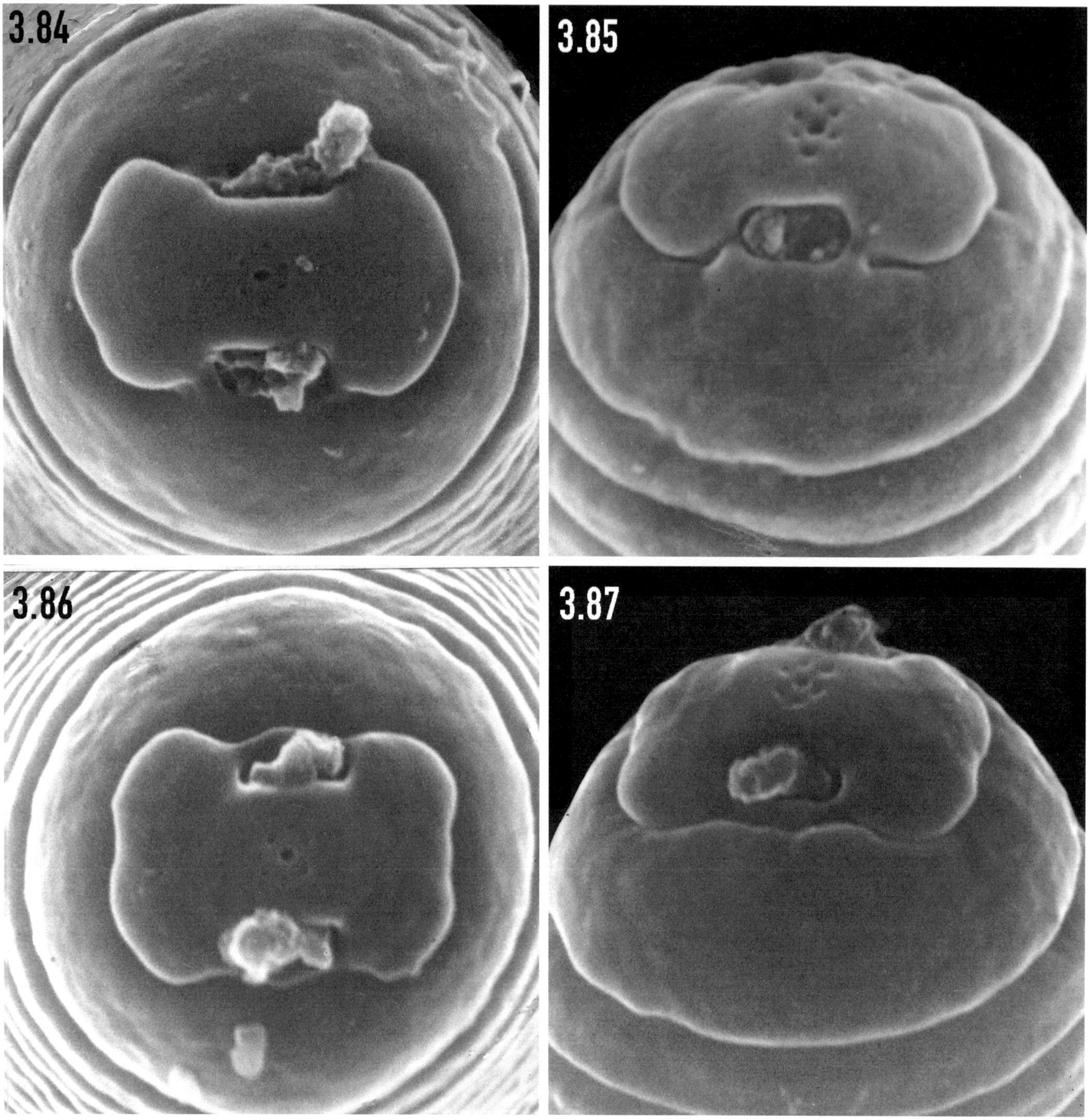

Figs. 3.84-3.87. Second-stage juveniles of *Meloidogyne hapla* race A (SEM). **3.84, 3.85.** Second-stage juvenile from a *M. hapla* population with 15 chromosomes, face and lateral view, respectively. **3.86, 3.87.** Second-stage juvenile from a *M. hapla* population with 16 chromosomes, face and lateral view, respectively

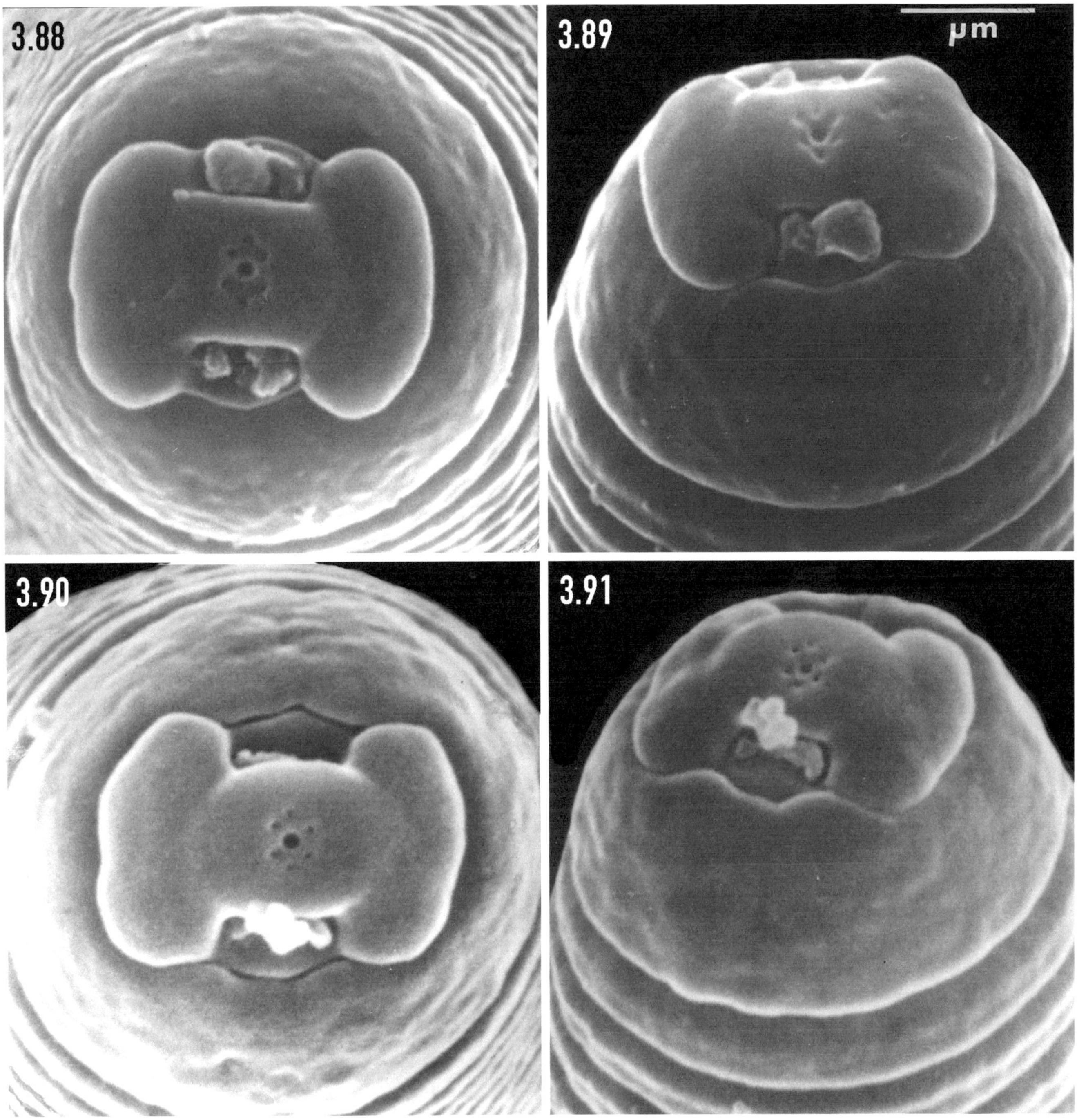

Figs. 3.88-3.91. Second-stage juveniles of *Meloidogyne hapla* (SEM). **3.88, 3.89.** *M. hapla* race A second-stage juvenile from a population with 15 chromosomes. **3.90, 3.91.** *Meloidogyne hapla* race B second-stage juvenile from a population with 48 chromosomes

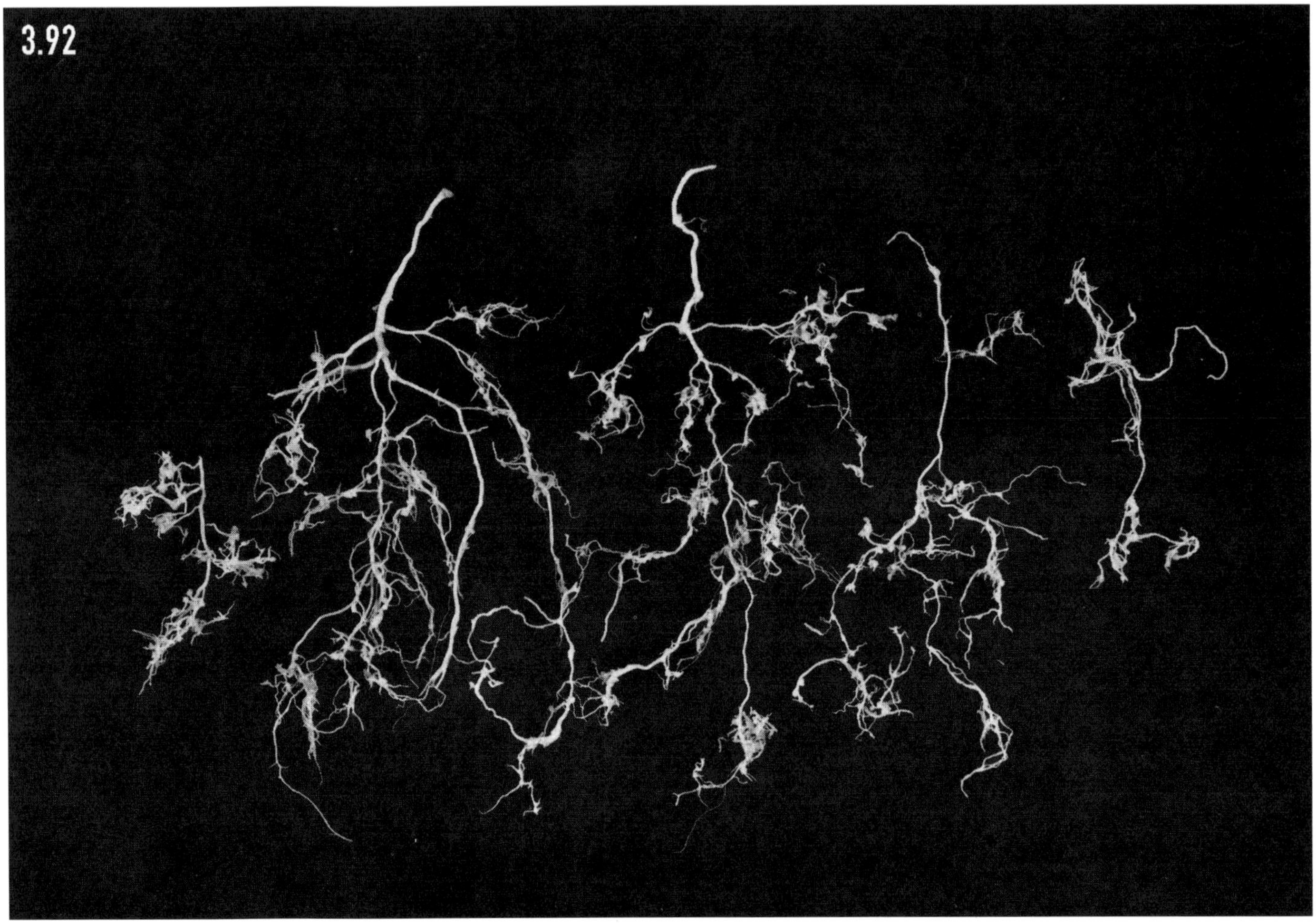

Fig. 3.92. Tomato roots with galls caused by *Meloidogyne hapla*

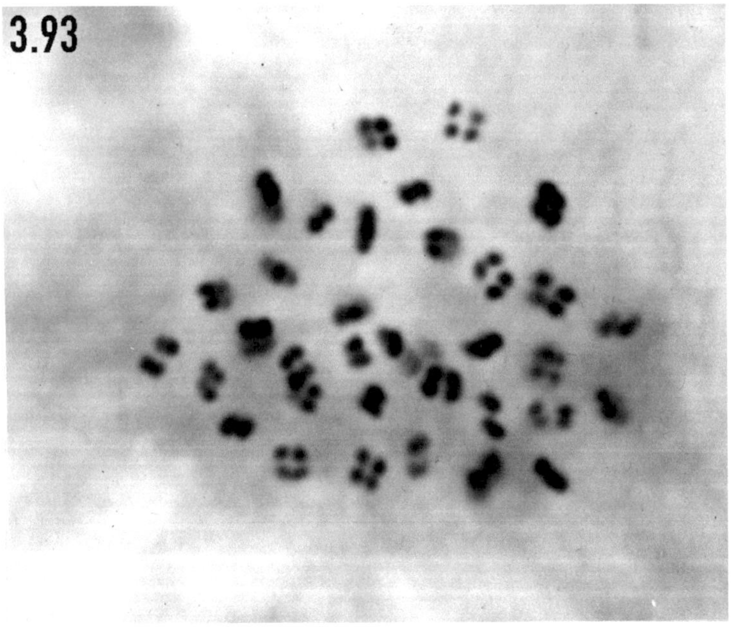

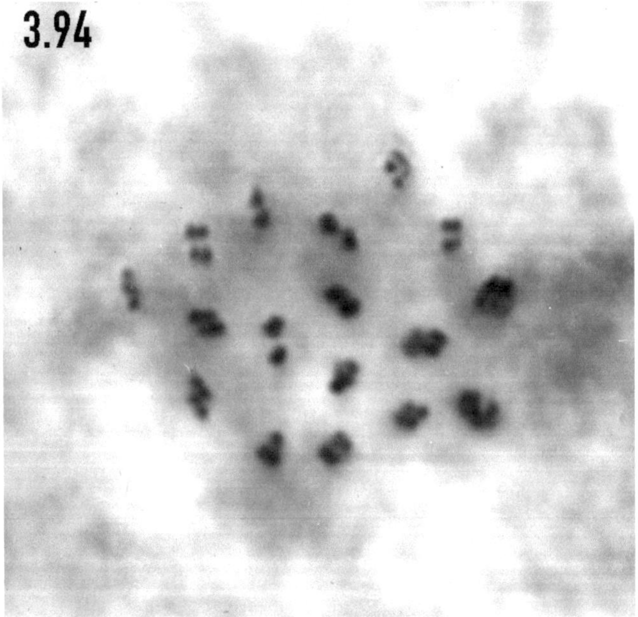

Figs. 3.93-3.94. Prometaphase chromosomes during the first maturation division of *M. hapla* race A. They are bivalent (tetrads) indicating that pairing of homologous chromosomes has occurred. The chromosomes of *M. hapla* race Bare similar to those of *M. arenaria* (After Triantaphyllou: J. Morphol. 118:403-414, 1966)

Appendix

Comparisons of the four common *Meloidogyne* species

I. Stylets of females

A. Comparison by LM (**Figs. A.1-A.4**) and SEM (**Figs. A.5-A.8**).

B. A key based on stylet morphology of females (**Table A.1**).

II. A key based on cytology (**Table A.2**).

M. incognita ***M. javanica*** ***M. arenaria*** ***M. hapla***

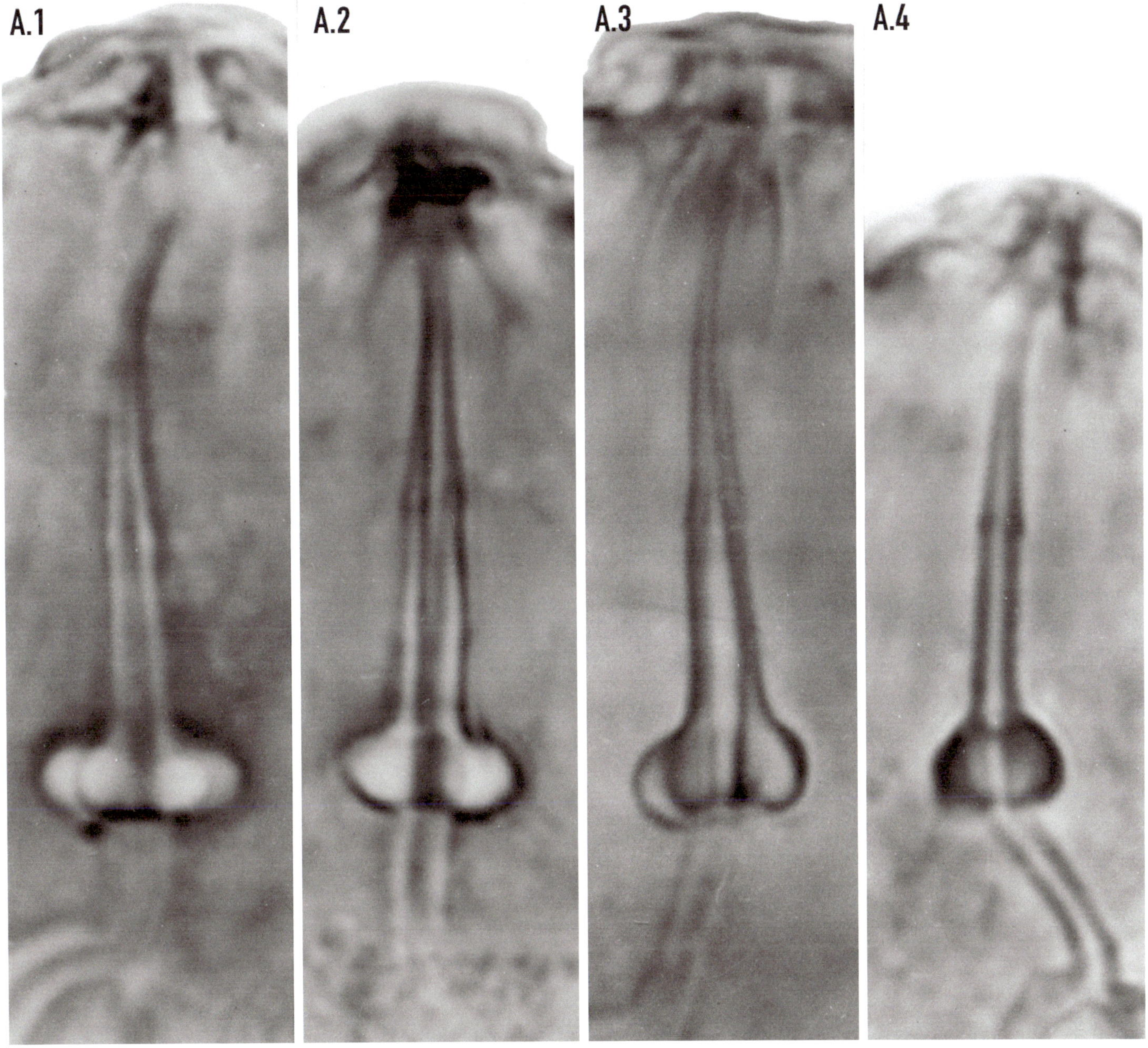

Figs. A.1-A.4. Comparison of stylets of females of *Meloidogyne incognita*, *M. javanica*, *M. arenaria*, and *M. hapla*, respectively (LM)

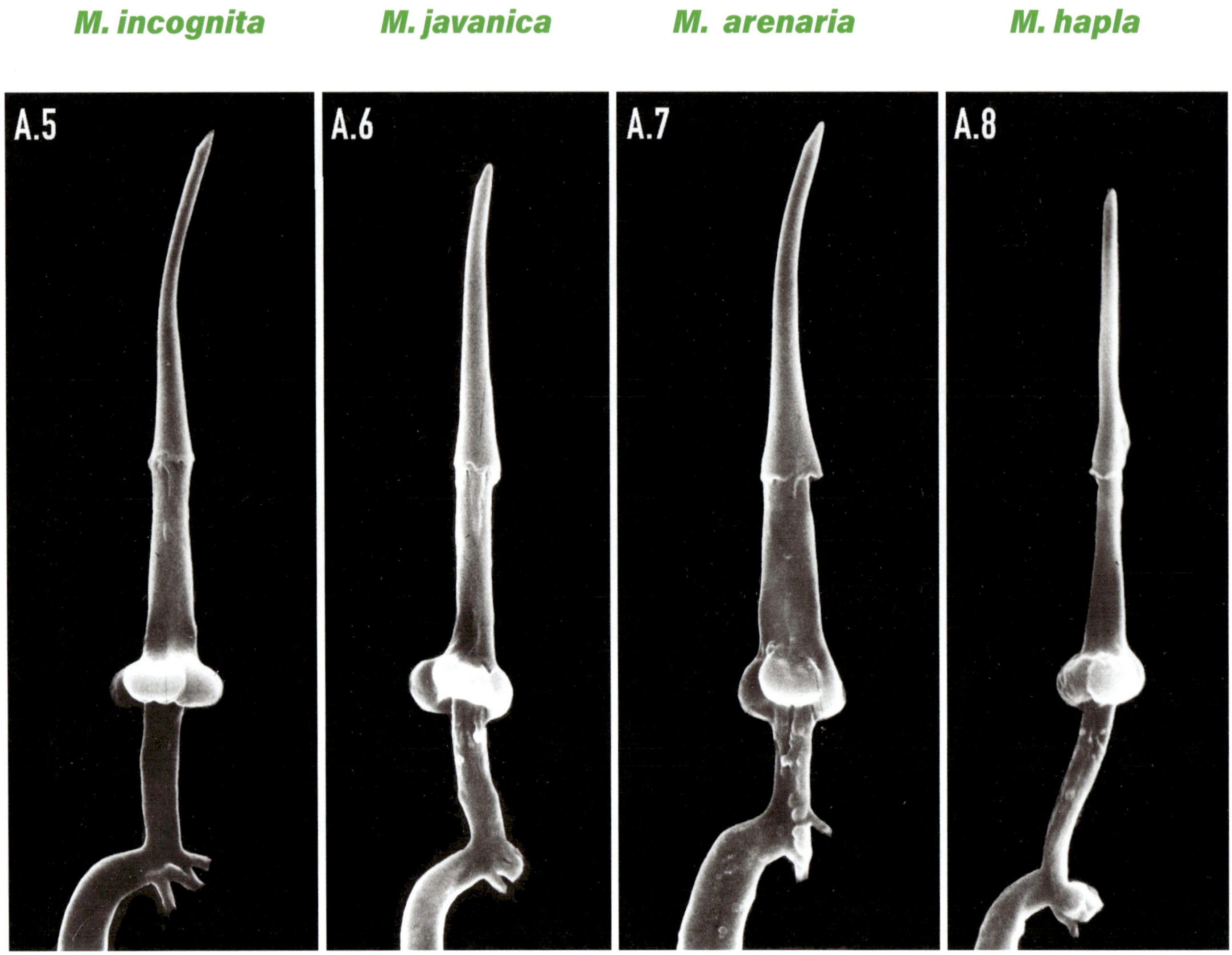

Figs. A.5-A.8. **Comparison of excised stylets of females of *Meloidogyne incognita, M. javanica, M. arenaria*, and *M. hapla*, respectively (SEM)**

Table A. I. **Key to the four most common *Meloidogyne* species based on stylet morphology of females.**

1.	Stylet delicate, knobs rounded and set off from shaft	*M. hapla*
	Stylet robust, knobs not rounded	2
2.	Stylet knobs gradually merge with shaft, entire stylet very broad	*M. arenaria*
	Stylet knobs broadly elongate and set off from shaft	3
3.	Anterior portion of cone distinctly curved dorsally	*M. incognita*
	Anterior portion of cone only slightly curved dorsally	*M. javanica*

Table A.2. Key to the most common *Meloidogyne* spp. based on cytological data

1. Prophase chromosomes nondiscrete, grouped together in a small area (**Figs. 3.17-3.18**). Most oocytes of a female are at prophase I, and only a few may have advanced to metaphase I........ .... *M. incognita*
   Prophase chromosomes well separated from each other (**Figs. 3.35, 3.53-3.54, 3.93-3.94**) Most oocytes of a female have advanced to anaphase and telophase I...2
2. Prometaphase and metaphase chromosomes are bivalent tetrads (**Figs. 3.93-3.94**) and in small numbers (14-19)...3
   Prometaphase and metaphase chromosomes are univalent dyads (**Figs. 3.35, 3.53-3.54**) and in large numbers (30-56)...4
3. Number of bivalent chromosomes n =19...*M. microtyla*
   Number of bivalent chromosomes n =18...Several species[1]
   Number of bivalent chromosomes n=17, 16, 15 or 14 (**Figs. 3.93-3.94**)............................*M. hapla* (race A)
4. Number of univalent chromosomes 2n =30-31 ....*M. hapla* (race B)
   Number of univalent chromosomes 2n =34-37...*M. arenaria* (race B)
   Number of univalent chromosomes 3n =42-48 (**Fig. 3.35**)..*M. javanica* or *M. hapla* (race B)[2]
   Number of univalent chromosomes 3n=50-56 (**Fig. 3.53**).................................*M. arenaria* (race A)

[1] In this group there can be any one of the following species: *M. carolinensis, M. megatyla, M. exigua, M. graminicola, M. naasi, M. graminis* or *M. ottersoni*. Also many other rare, described species, with limited host range and geographical distribution, are expected to belong to this category.

[2] Although there is an overlap in chromosome numbers between *M. javanica* and *M. hapla* (race B), the chances of error in identification are not great if morphological or host specificity characteristics are considered at the same time. Furthermore, the known geographical distribution is instructive. *M. javanica* is widely distributed in tropical, subtropical and some temperate regions. *M. hapla* (race B) with 43-45 chromosomes is very rare and is known only from a few populations in the USA and Europe. *M. hapla* (race B) with 48 chromosomes is known to occur only in Chile, S.A.

Selected References

Morphology

Chitwood, B. G. 1949. "Root-knot nematodes." I. A revision of the genus *Meloidogyne* Goeldi, 1887. Proceedings of the Helminthological Society of Washington 16:90-104.

Eisenback, J. D., and H. Hirschmann. 1979. Morphological comparison of second-stage juveniles of six populations of *Meloidogyne hapla* by SEM. Journal of Nematology 11:5-16.

Eisenback, J. D., and H. Hirschmann. 1979. Morphological comparison of second-stage juveniles of several *Meloidogyne* species (root-knot nematodes) by scanning electron microscopy. SEM/1979/III, SEM, Inc., AMF O'Hare, Ill. 60606. Pp. 223-230.

Eisenback, J. D., and H. Hirschmann. 1980. Morphological comparison of *Meloidogyne* males by scanning electron microscopy. Journal of Nematology 12:23-32.

Eisenback, J. D., H. Hirschmann, and A. C. Triantaphyllou. 1980. Morphological Comparison of *Meloidogyne* female head structures, perineal patterns, and stylets. Journal of Nematology 12:300-313.

Esser, R. P., V. G. Perry, and A. L. Taylor. 1976. A diagnostic compendium of the genus *Meloidogyne* (Nematoda: Heteroderidae). Proceedings of the Helminthological Society of Washington 43:138-150.

Franklin, M. T. 1965. *Meloidogyne*–Root-knot eelworms. Pp. 59-88, *in*, J. F. Southey, ed. Plant Nematology, Technical Bulletin No. 7. Her Majesty's Stationery Office, London.

Taylor, A. L., and J. N. Sasser. 1978. Biology, Identification and Control of Root-Knot Nematodes (*Meloidogyne* species). A Cooperative publication of the Department of Plant Pathology, North Carolina State University and the United States Agency for International Development. North Carolina State Graphics, Raleigh, N.C. 111 pp.

Taylor, A. L., V. H. Dropkin, and G. C. Martin. 1955. Perineal patterns of root-knot nematodes. Phytopathology 45:26-34.

Whitehead, A. G. 1968. Taxonomy of *Meloidogyne* (Nematodea: Heteroderidae) with descriptions of four new species. Transactions of the Zoological Society, London 31:263-401.

Differential Host Test

Taylor, A. L., and J. N. Sasser. 1978. Biology, Identification and Control of Root-Knot Nematodes (*Meloidogyne* species). A cooperative publication of the Department of Plant Pathology, North Carolina State University and the United States Agency for International Development. North Carolina State Graphics, Raleigh, N.C. 111 pp.

Cytogenetics

Triantaphyllou, A. C. 1979. Cytogenetics of root knot nematodes. Pp. 85-109. *in*, Lamberti, F., and C. E. Taylor, eds. Root-knot nematodes (*Meloidogyne* Species). Systematics, Biology and Control. Academic Press, New York.

Triantaphyllou, A. C., and H. Hirschmann. 1980. Cytogenetics and morphology in relation to evolution and speciation of plant-parasitic nematodes. Annual Review of Phytopathology 18:333-359.

Biochemistry

Dalmasso, A., and J. B. Berge. 1978. Molecular polymorphism and phylogenetic relationship in some *Meloidogyne* spp: Application to the taxonomy of *Meloidogyne*. Journal of Nematology 10:323-332.

Hussey, R. 1979. Biochemical systematics of nematodes - A review. Helminthological Abstracts 48:141-148.

Ecology

Sasser, J. N., A. L. Taylor, and L. A. Nelson. 1980. Ecological factors influencing survival and pathogenicity of *Meloidogyne* species. Journal of Nematology 12:237 (Abstr.).

Van Gundy, S. D. 1976. Ecology of root-knot nematodes. Pp. 64-74, *in*, Proceedings of the research planning conference on root-knot nematodes *Meloidogyne*, spp. North Carolina State University, Raleigh, North Carolina, January 12-16, 1976.

Made in the USA
Las Vegas, NV
03 November 2026

57423722R00036